Water Supply System Analysis - Selected Topics

Water Supply System Analysis - Selected Topics

Editor

Keith Marcell

scitus academics

Water Supply System Analysis - Selected Topics
Edited by **Keith Marcell**

ISBN: 978-1-68117-620-8
Library of Congress Control Number: 2016946478

SCITUS Academics LLC,
616, Corporate Way, Suite 2, 4766,
Valley Cottage, NY 10989

www.scitusacademics.com

Notice

Printed in United States of America on Acid Free Paper

Preface

Water supply system is an infrastructure for the distribution, storage, treatment, transmission and collection of water for homes, commercial establishments, industry and irrigation, as well as for good public needs such as firefighting and street flushing. People rely on water to drink, cook, wash, carry waste and other household needs. Water supply systems also have to comply with public, professional and manufacturing requirements. Many attempts have been made to develop a water supply system for a sustainable water supply. However, in the very first era the complexity of the system limited the specific application of the site. As water demands on the current water supply system increase more and more, numerous studies have attempted to produce a wide water supply system to help decision-makers design more efficient systems for a certain range of operations. These efforts include optimizing the total construction and operating costs of the system. Water has to meet both quantity and quality requirements in most cases.

This book covers theories and practical models for the analysis of water supply systems, including: transient investigation strategy, sustainable management of local water supply networks, infrastructure management, and uncertainty of demand, water meter measurement errors and water indicators.

Table of Contents

Chapter 1

Guidelines for Transient Analysis in Water Transmission and Distribution Systems

Ivo Pothof[1, 2] and Bryan Karney[3]

[1] Deltares, MH Delft, The Netherlands

[2] Delft University of Technology, Department of Water Management, Stevinweg, CN Delft, The Netherlands

[3] University of Toronto, Canada and HydraTek and Associates Inc., Canada

1. INTRODUCTION

Despite the addition of chlorine and potential flooding damage, drinking water is not generally considered a hazardous commodity nor an overwhelming cost. Therefore, considerable water losses are tolerated by water companies throughout the world. However, more extreme variations in dry and wet periods induced by climate change will demand more sustainable water resource management. Transient phenomena ("transients") in water supply systems (WSS), including transmission and distribution systems, contribute to the occurrence of leaks. Transients are caused by the normal variation in drinking water demand patterns that trigger pump operations and valve manipulations. Other transients are categorised as incidental or emergency operations. These include events like a pumping station power failure or an accidental pipe rupture by external forces. A number of excellent books on fluid transients have been written (Tullis 1989; Streeter and Wylie 1993; Thorley 2004), which focus on the physical phenomena, anti-surge devices and numerical modelling.

However, there is still a need for practical guidance on the hydraulic analysis of municipal water systems in order to reduce or counteract the adverse effects of transient pressures. The need for guidelines on pressure transients is not only due to its positive effect on water losses, but also by the contribution to safe, cost-effective and energy-saving operation of water distribution systems. This chapter addresses the gap of practical guidance on the analysis of pressure transients in municipal water systems.

All existing design guidelines for pipeline systems aim for a final design that reliably resists all "reasonably possible" combinations of loads. System strength (or resistance) must sufficiently exceed the effect of system loads. The strength and load evaluation may be based on the more traditional allowable stress approach or on the more novel reliability-based limit state design. Both approaches and all standards lack a methodology to account for dynamic hydraulic loads (i.e., pressure transients) (Pothof 1999; Pothof and McNulty 2001). Most of the current standards simply state that dynamic internal pressures should not exceed the design pressure with a certain factor, duration and occurrence frequency. The Dutch standard NEN 3650 (Requirements for pipeline systems) includes an appendix that provides some guidance on pressure transients (NEN 2012).

One of the earliest serious contributions to this topic was the significant compilation of Pejovic and Boldy (1992). This work not only considered transient issues such as parameter sensitivity and data requirements, but usefully classified a range of loading conditions that accounted for important differences between normal, emergency and catastrophic cases, and the variation in risk and damage that could be tolerated under these different states.

Boulos *et al.* (2005) introduced a flow chart for surge design in WSS. The authors address a number of consequences of hydraulic transients, including maximum pressure, vacuum conditions, cavitation, vibrations and risk of contamination. They proposed three potential solutions in case the transient analysis revealed unacceptable incidental pressures:

1) Modification of transient event, such as slower valve closure or a flywheel;

2) Modification of the system, including other pipe material, other pipe routing, etc.; and

3) Application of anti-surge devices.

Boulos *et al.* list eight devices and summarise their principal operation. They do not provide an overview of the scenarios that should be included in a pressure transient analysis. Jung and Karney (2009) have recognised that an *a priori* defined design load does not necessarily result in the worst-case transient loading. Only in very simple systems can the most critical parameter combination can be defined *a priori* (Table 4). In reality, selecting appropriate boundary conditions and parameters is difficult. Further, the search for the worst case scenario, considering the dynamic behaviour in a WSS, is itself a challenging task due to the complicated nonlinear interactions among system components and variables. Jung and Karney (2009) have extended the flow chart of Boulos *et al.* (2005), taking into account a search for the worst-case scenario (Figure 1). They propose to apply optimisation tools to find the worst-case loading and a feasible set of surge protection devices.

Automatic control systems have become common practice in WSS. Since WSS are spatially distributed, local control systems may continue in normal operating mode, after a power failure has occurred somewhere else in the system. The control systems may have a positive or negative effect on the propagation of hydraulic transients. On the other hand, the distributed nature of WSS and the presence of control systems may be exploited to counteract the negative effects of emergency scenarios. Therefore, existing guidelines on the design of WSS must be updated on a regular basis in order to take these developments into account.

Typical design criteria for drinking water and wastewater pipeline systems are listed in section 2. Section 3 presents a systematic approach to the surge analysis of water systems. This approach focuses on guidelines for practitioners. The key steps in the approach include the following: preconditions for the surge analysis; surge analysis of emergency scenarios without provisions; sizing of anti-surge provisions and design of emergency controls; evaluation of normal operations and design of control systems. The approach has been applied successfully by Both Deltares (formerly Delft Hydraulics) and HydraTek and Associates Inc. in numerous large water transmission schemes worldwide. Especially the integrated design of surge provisions and control systems has many benefits for a safe, cost-effective and energy-efficient operation of the water pipeline system. Section 4 summarises the key points of this paper.

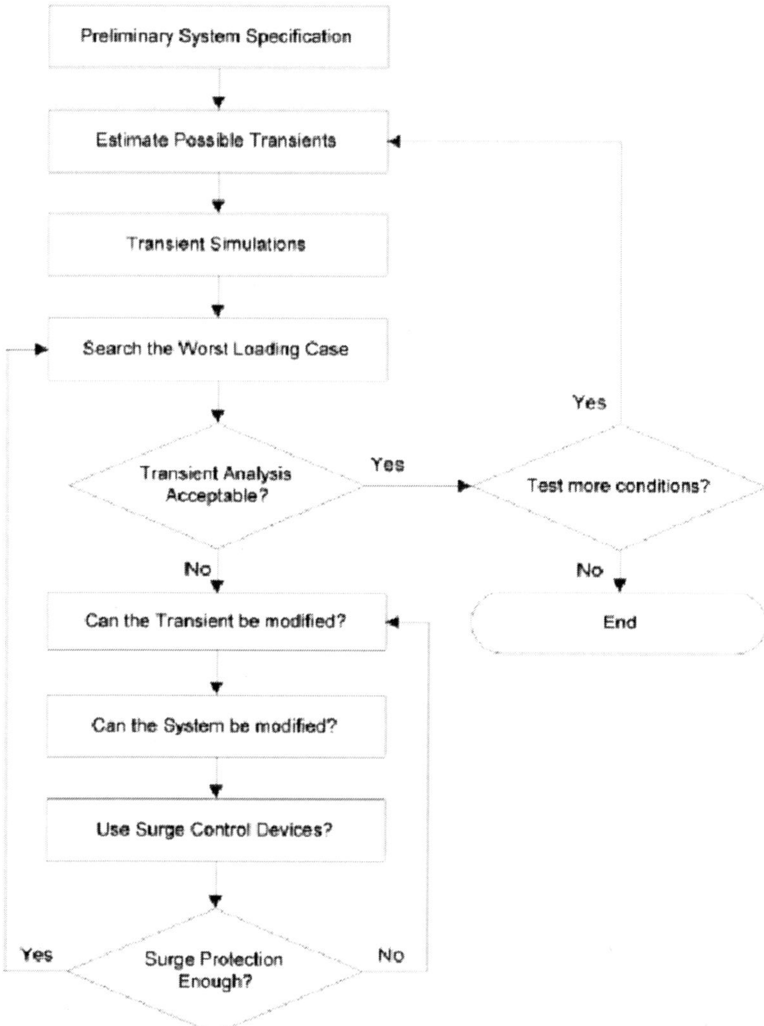

Figure 1. Pressure Transient design (**Jung and Karney 2009**).

2. PRESSURE TRANSIENT EVALUATION CRITERIA FOR WATER PIPELINES

In any transient evaluation, pressure is the most important evaluation variable, but certainly not the only one. Component-specific criteria must be taken into account as well, such as a minimum fluid level in air vessels, maximum air pressure during air release from an air valve or the maximum fluid deceleration through an undamped check valve.

The maximum and minimum allowable pressure is directly related to the pressure rating of the components. Thin-walled steel and plastic pipes are susceptible to buckling at a combination of external pressure and minimum internal pressure.

The design pressure for continuous operation is normally equal to the pressure rating of the system. During transient events or emergency operation, the system pressure may exceed the design pressure up to a certain factor of the design pressure. Table 1 provides an overview of maximum allowable incidental pressure (MAIP) in different national and international codes and standards.

Table 1. Overview of maximum allowable incidental pressures (MAIP) in international standards, expressed as a factor of the nominal pressure class.

Code	Maximum Incidental Pressure Factor [-]
DVGW W303:1994 (German guideline)	1.00
ASME B31.4 (1992), IS 328, BS 8010, ISO CD 16708:2000	1.10
NEN 3650-1:2012	1.15
BS 806	1.20
Italian ministerial publication	1.25 – 1.50

The minimum allowable pressure is rarely explicitly addressed in existing standards. The commonly accepted minimum incidental pressure in drinking water distribution systems is atmospheric pressure or the maximum groundwater pressure necessary to avoid intrusion at small leaks. If the water is not for direct consumption, negative pressures down to full vacuum may be allowed if the pipe strength is sufficient to withstand this condition, although tolerance to such conditions varies with jurisdiction. Full vacuum and cavitation can be admitted under the condition that the cavity implosion is admissible. Computer codes that are validated for cavity implosion must be used to determine the implosion shock. The maximum allowable shock pressure is 50% of the design pressure. This criterion is based on the following reasoning: The pipeline (including supports) is considered a single-mass-spring system for which a simplified structural dynamics analysis can be carried out. The ratio of the dynamic response (i.e., pipe wall stress) to the static response is called the dynamic load factor (DLF). The dynamic load factor of a mass-spring system is equal to 2. It is therefore recommended that a maximum shock pressure of no more than 50% of the design pressure be allowed. This criterion may be relaxed if a more complete Fluid-Structure-Interaction (FSI) simulation is performed for critical above-ground pipe sections.

3. SYSTEMATIC APPROACH TO PRESSURE TRANSIENT ANALYSIS

The flow chart in Figure 2 integrates the design of anti-surge devices and distributed control systems. It is emphasised that a surge analysis is strongly recommended upon each modification to an existing system. The systematic approach also applies to existing systems.

Because system components are tightly coupled, detailed economic analysis can be a complex undertaking, However, the net present value of anti-surge equipment may rise to 25% of the total costs of a particular system. Therefore, the systematic approach to the pressure transient analysis is preferably included in a life cycle cost optimisation of the water system, because savings on investment costs may lead to operation and maintenance costs that exceed the net present value of the investment savings.

3.1. Necessary Information For A Pressure Transient Analysis

The phenomenon of pressure transients, surge or water hammer is defined as the simultaneous occurrence of a pressure and velocity changes in a closed conduit. Water hammer may occur in both long and short pipes. The larger and faster the change of velocity, the larger the pressure changes will be. In this case, 'fast' is not an absolutely term, but can only be used relative to the pipe period, that is, relative to the pipe's internal communications. The most important parameters for the magnitude of transient pressures are:

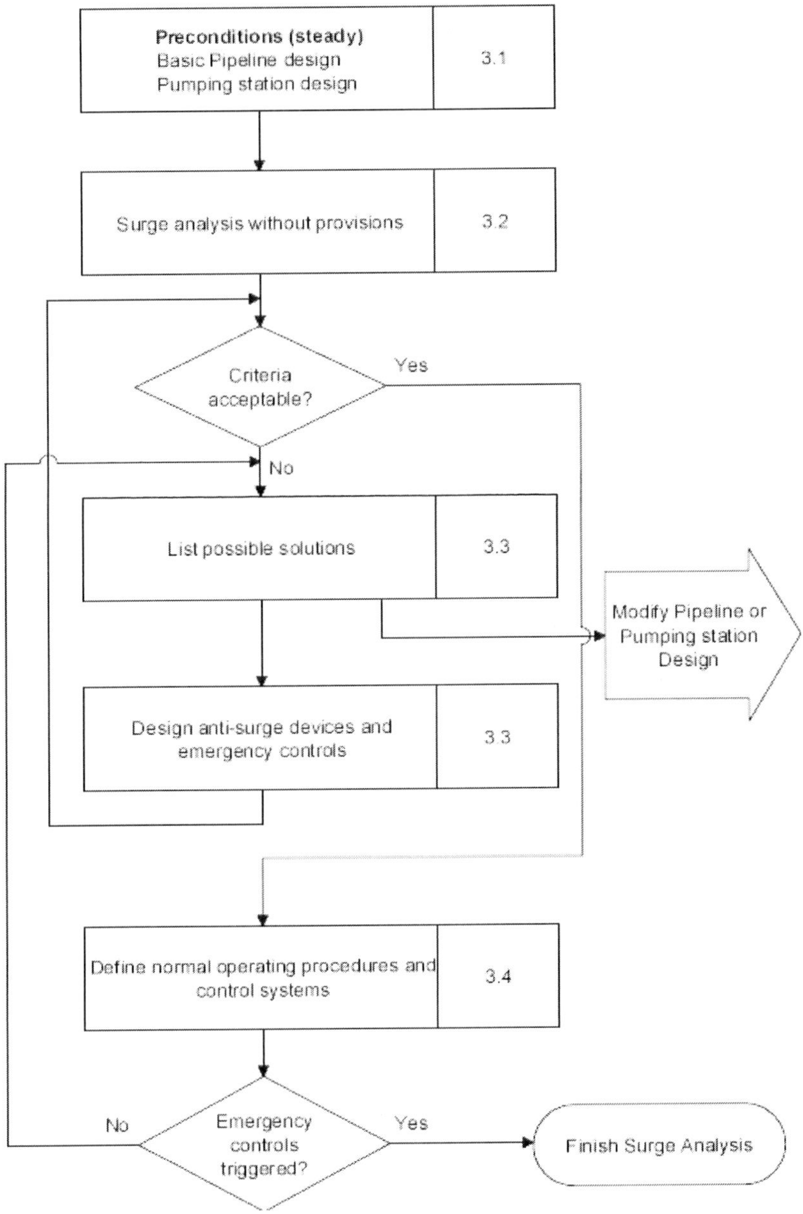

Figure 2. Integrated design for pressure transients and controls.

- Velocity change in time, Δv (m/s) (or possibly the pressure equivalent)
- Acoustic wave speed, c (m/s)
- Pipe period, T (s)
- Joukowsky pressure, Δp (Pa)
- Elevation profile

The acoustic wave speed c is the celerity at which pressure waves travel through pressurised pipes. The wave speed accounts for both fluid compressibility and pipe stiffness: the more elastic the pipe, the lower the wave speed. In fact, all phenomena that create internal storage contribute to a reduction of wave speed. Since air is much more compressible than water, air bubbles reduce the wave speed considerably, but this is the primary positive effect of air in pipelines. The negative consequences of air in water pipelines, particularly in permitting or generating large velocity changes, can greatly exceed this positive effect in mitigating certain transient changes; thus, as an excellent precaution, free or mobile air must generally be avoided in water systems whenever possible and cost-effective. The maximum acoustic wave speed in an excavated water tunnel through rocks is 1430 m/s and drops to approximately 1250 m/s in steel, 1000 m/s in concrete and ductile iron, 600 m/s in GRP, 400 m/s in PVC and about 200 m/s in PE pipes.

$$c = \frac{1}{\sqrt{\rho\left(\frac{C_1 D}{eE} + \frac{1}{K}\right)}} \tag{1}$$

where:

c = Acoustic wave speed (m/s)

E = Young's modulus of pipe material (N/m^2)

K = Bulk modulus of fluid (N/m^2)

ρ = Fluid density (kg/m^3)

D = Pipe diameter (m)

e = Wall thickness (m) and

C_1 = Constant depending on the pipe anchorage (order 1).

The acoustic wave speed in water pipelines is shown in Figure 3.

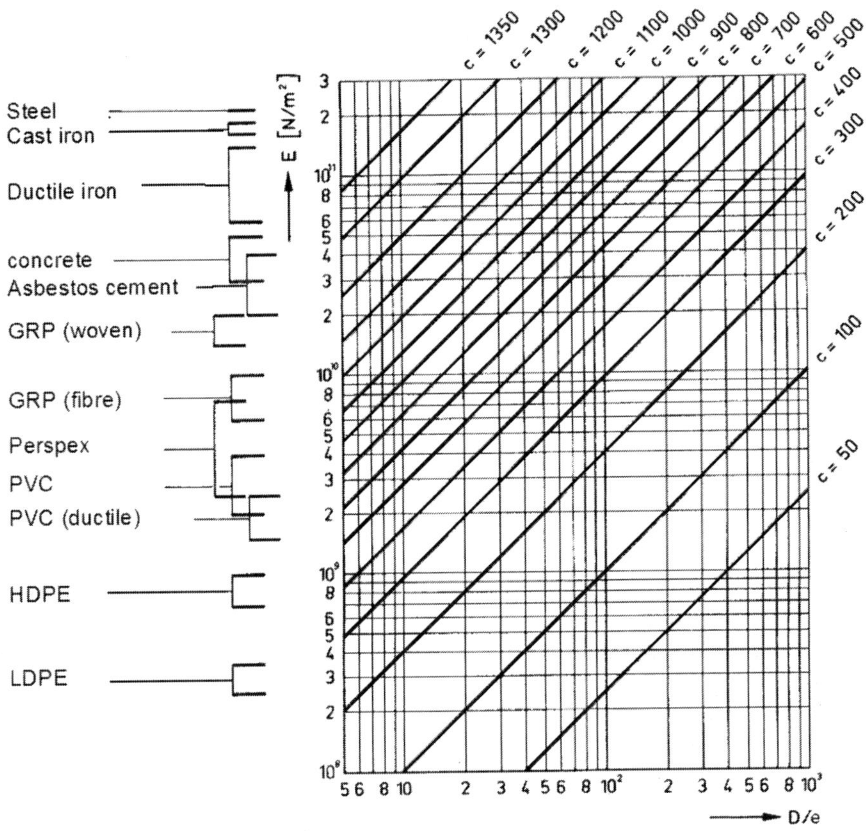

Figure 3. Graph of acoustic wave speed in water pipelines in relation to pipe material (E) and wall thickness (D/e).

The pipe period *T* [s] is defined as the time required for a pressure wave to travel from its source of origin through the system and back to its source. For a single pipeline with length L:

$$T = 2L/c \tag{2}$$

This parameter defines the natural time scale for velocity and pressure adjustments in the system.

Only after the pipe period the pressure wave will start to interact with other pressure waves from the boundary condition, such as a tripping pump or a valve closure. Any velocity change Δv within the pipe period will result in a certain "practical maximum" pressure, the so-called Joukowsky pressure,Δp.

$$\Delta p = \pm \rho \cdot c \cdot \Delta v \tag{3}$$

A slightly more conservative assessment of the maximum transient pressure includes the steady friction head loss $\Delta ps = \rho g \Delta Hs$.

$$\Delta p = \pm \left(\rho \cdot c \cdot \Delta v + \rho g \Delta H_s \right) \tag{4}$$

All these parameters follow directly from the basic design. The maximum rate of change in velocity is determined by the run-down time of a pump or a valve closure speed. The pump run-down time is influenced by the polar moment of inertia of the pump impeller, the gear box and motor. The full stroke closure time of valves may be increased in order to reduce the rate of velocity change.

Pressure waves reflect on variations of cross-sectional area (T-junctions, diameter changes, etc.) and variation of pipe material. All these parameters must be included in a hydraulic model.

Finally, the elevation profile is an important input, because extreme pressures typically occur at its minimum and maximum positions.

3.2. Emergency Scenarios Without Anti-Surge Provisions

A pressure transient analysis or surge analysis includes a number of simulations of emergency scenarios, normal operations maintenance procedures. The emergency scenarios may include:

- Complete pump trip
- Single pump trip to determine check valve requirements
- Unintended valve closure; and
- Emergency shut-down procedures.

A pump trip without anti-surge provisions causes a negative pressure wave traveling into the WSS. If the downstream boundary is a tank farm or large distribution network, then the reflected pressure wave is an overpressure wave. If the check valves have closed within the pipe period, then the positive pressure reflects on the closed check valves by doubling the positive pressure wave (Figure 4). In this way, the maximum allowable pressure may be exceeded during a pump trip scenario.

Check valves will generally close after pump trip. The transient closure of a check valve is driven by the fluid deceleration through the check valve. If the fluid decelerates quickly, an undamped check valve will slam in reverse flow. Fast-closing undamped check valves, like a nozzle- or piston-type check valve, are designed to close at a very small return velocity in order to minimize the shock pressure. Ball check valves are relatively slow, so that their application is limited to situations with small fluid decelerations.

Emergency closure of a line valve creates a positive pressure wave upstream and negative pressure wave downstream of the valve. Although the total closure time may well exceed the characteristic pipe period, the effective closure may still occur within one pipe period, so that the Joukowsky pressure shock may still occur. The effective closure is typically only 20% of the full stroke closure time, because the valve starts dominating the total head loss when the valve position is less than 20% open (e.g., Figure 6). If a measured capacity curve of the valve is used, simulation software will deliver a reliable evolution of the discharge and transient pressures in the WSS.

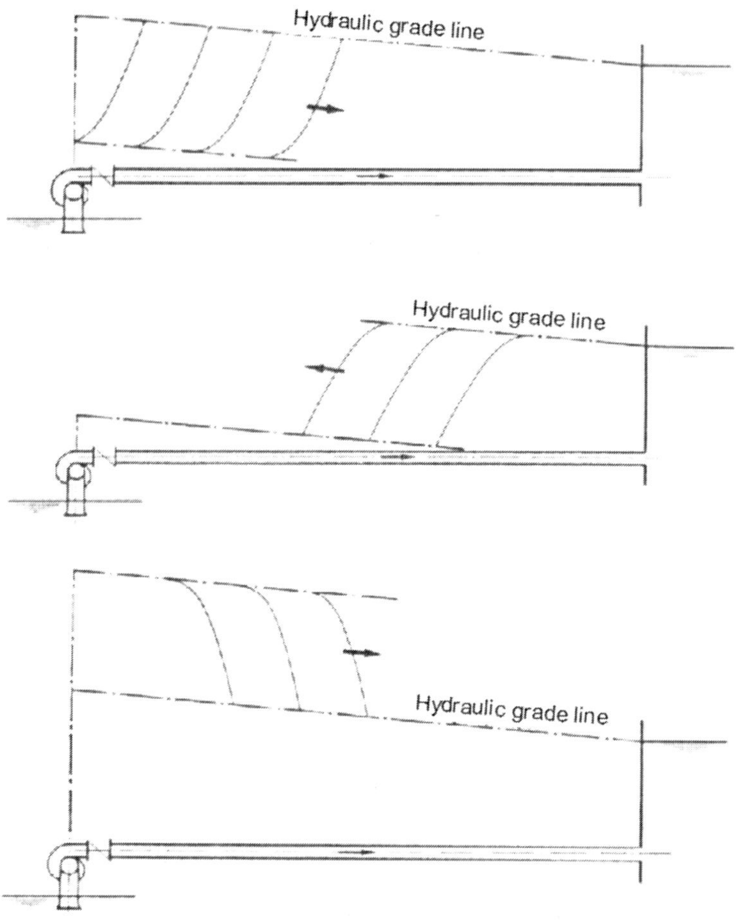

Figure 4. Pressure wave propagation following a pump trip

Figure 6 shows an example of a butterfly valve at the end of a 10 km supply line (wave speed is 1000 m/s). A linear closure in 5 pipe periods (100 s) shows that the pressure rises only during the last 30% of the valve closure. Therefore the pressure rise is almost equal to the Joukowsky pressure. A two-stage closure, with a valve stroke from 100% to 30% open in 1 pipe period (20 s), shows a more gradual pressure rise during the closing procedure and a lower peak pressure.

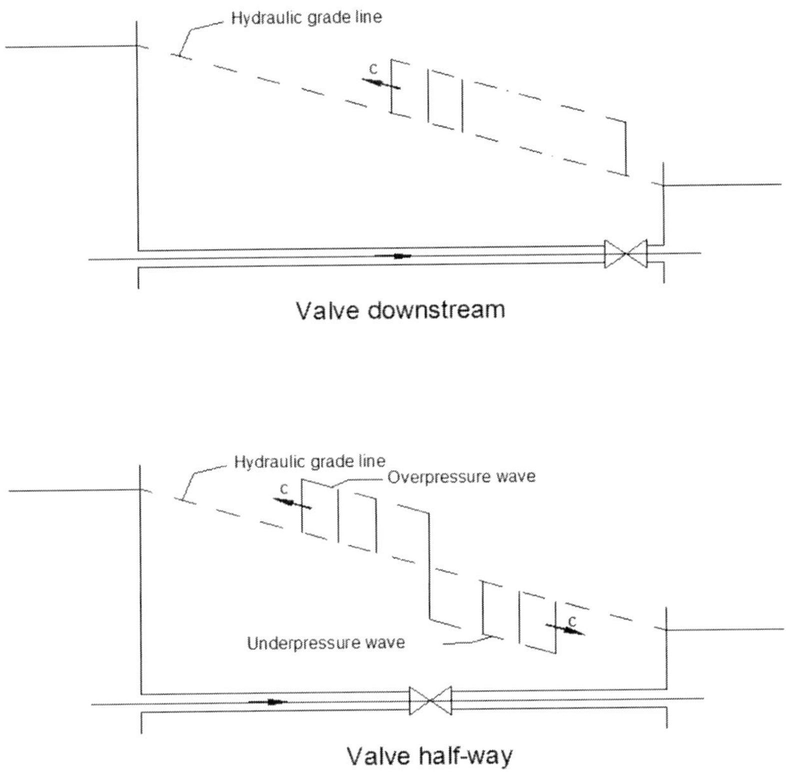

Figure 5. Pressure wave propagation following valve closure

In general, for each scenario multiple simulations must be carried out to determine the extreme pressures and other hydraulic criteria. Scenario variations may include flow distributions, availability of signal transfer (wireless or fiber-optic cable) for the control system and parameter variations. For example, the minimum pressure upon full pump trip will be reached in a single pipeline, if the maximum wall roughness value is used. If an air vessel is used as an anti-surge device, the minimum wall roughness and isothermal expansion must be applied to determine the minimum water level in the air vessel. Adiabatic pocket expansion in air vessels must be applied for other scenarios. The selection of input parameters so that the extreme hydraulic criterion values are computed is called a conservative modeling approach (Pothof and McNulty 2001). The proper combination of input parameters can be determined *a priori* for simple (single pipeline) systems only. Table 4 provides some guidance on the conservative modeling approach.

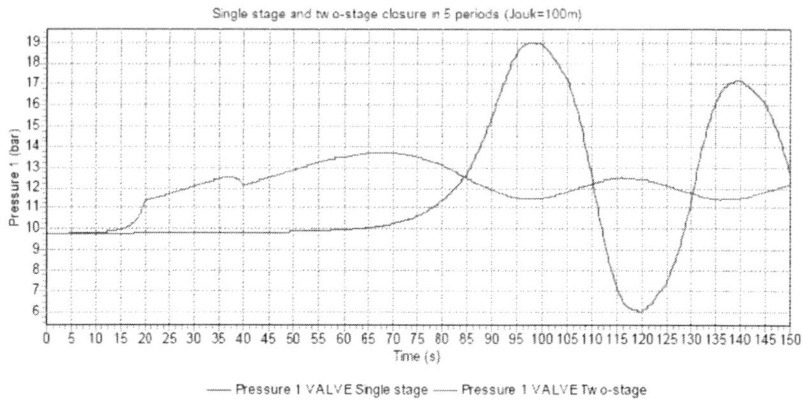

Figure 6. Single and two-stage valve in 5 pipe periods (100 s)

In more realistic situations a sensitivity analysis is required to determine the worst case loading. A more recent development for complex systems is to combine transient solvers with optimization algorithms to find the worst case loading condition and the appropriate protection against it (Jung and Karney 2009).

In most cases, the emergency scenarios result in inadmissible transient pressures. Possible solutions include modifications to the system or transient event (e.g., slower valve closure), anti-surge devices, emergency controls, or a combination of the above. The solutions will be discussed in more detail in the next section.

3.3. Design Of Anti-Surge Devices And Emergency Controls

In order to mitigate inadmissible transient pressures, hydraulic design engineers have four different management options at their disposal:

1) System modifications (diameter, pipe material, elevation profile, etc.);

2) Moderation of the transient initiation event;

3) Emergency control procedures; and/or

4) Anti-surge devices.

3.3.1. System Modifications

Measure 1 is only feasible in an early stage. A preliminary surge analysis may identify cost-effective measures for the surge protection that cannot later be incorporated. If, for example, inadmissible pressures occur at a local high point that seem difficult to mitigate, the pipe routing may be changed to avoid the high point. Alternatively, the pipe may be drilled through a slope to lower the maximum elevation.

Selection of a more flexible pipe material reduces the acoustic wave speed. Larger diameters reduce the velocities and velocity changes, but the residence time increases, which may render this option infeasible due to quality concerns.

A cost-benefit analysis is recommended to evaluate the feasibility of these kinds of options.

3.3.2. Moderating The Transient Initiation Event

A reduction of the rate of velocity change will reduce the transient pressure amplitude. A variable speed drive or soft start/stop functionality may be effective measures for normal operations, but their effect is negligible in case of a power failure. A flywheel increases the polar moment of inertia and thereby slows down the pump trip response. It should be verified that the pump motor is capable of handling the large inertia of the flywheel during pump start scenarios. Experience shows that a flywheel is not a cost-effective option for pumps that need to start and stop frequently.

If inadmissible pressures are caused by valve manipulations, the valve closure time must be increased. The velocity reduction by a closing valve is not only influenced by the valve characteristic, but also by the system. The valve resistance must dominate the total system resistance before the discharge is significantly reduced. Therefore, the effective valve closure time is typically 20% to 30% of the total closure time. A two-stage closure, or the utilization of a smaller valve in parallel, may permit a rapid initial stage and very slow final stage as an effective strategy for an emergency shut down scenario. The effective valve closure must be spread over multiple pipe periods to obtain a significant reduction of the peak pressure. Existing books on fluid transient provide more detail on efficient valve stroking (Tullis 1989; Streeter and Wylie 1993; Thorley 2004).

3.3.3. Emergency Control Procedures

Since WSS are spatially distributed, the power supply of valves and pumps in different parts of the system is delivered by a nearly-independent power supply. Therefore, local control systems may continue operating normally, after a power failure has occurred somewhere else in the network. The control systems may have a positive or negative effect on the propagation of hydraulic transients. The distributed nature of WSS and the presence of control systems may be exploited to counteract the negative effects of emergency scenarios.

If a centralised control system is available, valves may start closing or other pumps may ramp up as soon as a pump trip is detected. Even without a centralised control system, emergency control rules may be developed to detect power failures. These emergency control rules should be defined in such a way that false triggers are avoided during normal operations. An example of an emergency control rule is: *ESD valve closure is initiated if the discharge drops by more than 10% of the design discharge and the upstream pressure falls by at least 0.5 bar within 60 seconds.*

3.3.4. Anti-Surge Devices

The above-described measures may be combined with one or more of the following anti-surge devices in municipal water systems.

An important distinction is made in Table 2 between anti-surge devices that directly affect the rate of change in velocity and anti-surge devices that are activated at a certain condition. The anti-surge devices in the first category immediately affect the system response; they have an overall impact on system behaviour. The pressure-limiting devices generally have a local impact. Table 3 lists possible measures when certain performance criteria are violated.

Table 2. Summary of anti-surge devices

Devices, affecting velocity change in time	Pressure limiting devices
Surge vessel	By-pass check valve
Flywheel	Pressure relief valve
Surge tower	Combination air/vacuum valves
	Feed tank

The surge vessel is an effective (though relatively expensive) measure to protect the system downstream of the surge vessel against excessive transients. However, the hydraulic loads in the sub-system between suction tanks and the surge vessel will increase with the installation of a surge vessel. Special attention must be paid to the check valve requirements, because the fluid deceleration may lead to check valve slam and consequent damage. These local effects, caused by the installation of a surge vessel, should always be investigated in a detailed hydraulic model of the subsystem between tanks and surge vessels. This model may also reveal inadmissible pressures or anchor forces in the suction lines and headers, especially in systems with long suction lines (> 500 m). A sometimes-effective measure to reduce the local transients in the pumping station is to install the surge vessels at a certain distance from the pumping station.

One of the disadvantages of a surge tower is its height (and thus cost and the siting challenges). If the capacity increases, so that the discharge head exceeds the surge tower level, then the surge tower cannot be used anymore. A surge tower is typically installed in the vicinity of a pumping station in order to protect the WSS downstream. A surge tower could also be installed upstream of a valve station to slow down the over pressure due to an emergency valve closure.

Table 3. Possible mitigating measures in case of violation of one or more performance criteria

Operation	Criterion Violation	Improvement
pump trip	low pressure	bypass pipe, flywheel larger pipe diameter air vessel, accumulator surge tower, surge vessel, feed tank air valve(s) at low pressure points in the system other pipe material with lower Young's modulus
pump trip	high pressure	air vessel with check valve and throttled by-pass
pump trip	reverse flow in pump	increase (check) valve closure rate by choosing an appropriate fast-closing check valve (e.g. nozzle type)
pump trip	rate of fluid deceleration through check valve (high pressure due to valve closure)	apply spring to reduce check valve closing time apply spring or counter weight with damper to increase check valve closing time and allow return flow
valve closure	high pressure (upstream)	air vessel slower valve closure pressure relief valve or damper at high pressure points higher pressure rating
valve closure	low pressure (downstream)	air vessel slower valve closure air valves at low pressure points
valve throttling	pressure instability	use multiple valves adjust control settings
drainage, filling	entrapped air	use air valves prevent drainage on shut-down

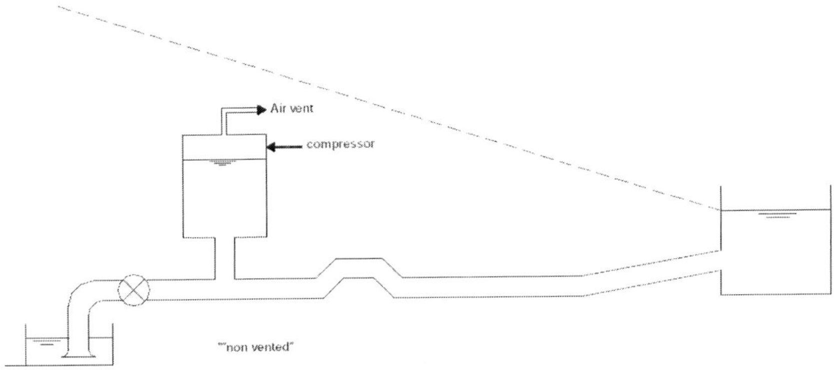

Figure 7. Non-aerated surge vessel

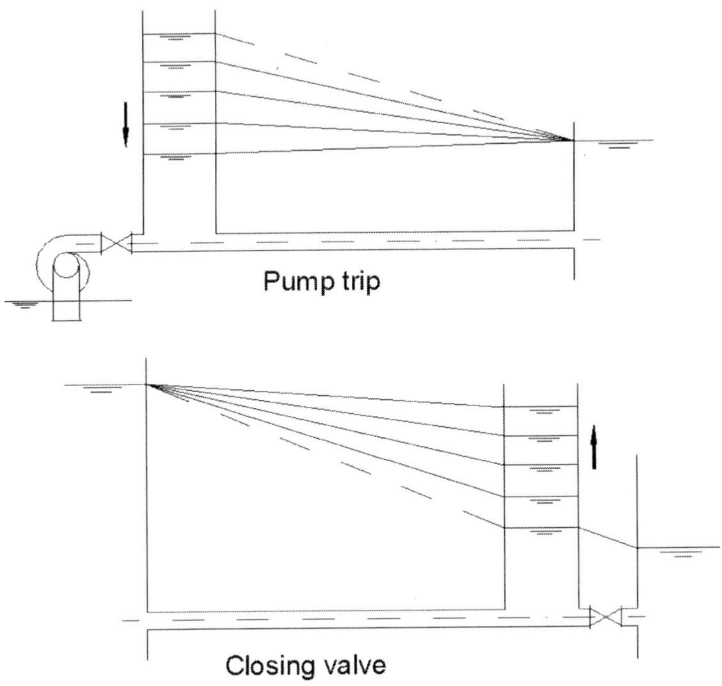

Figure 8. Surge tower near pumping station or valve station.

Another device that reduces the velocity change in time is the flywheel. A flywheel may be an effective measure for relatively short transmission lines connected to a tank farm or distribution network. A flywheel can be an attractive measure if the following conditions are met:

1) Pump speed variations are limited.

2) The pump motor can cope with the flywheel during pump start-up, which means that the motor is strong enough to accelerate the pump impeller - flywheel combination to the pump's rated speed. If the polar moment of pump and flywheel inertia is too large for the motor, then a motor-powered trip may occur and the rated speed cannot be reached.

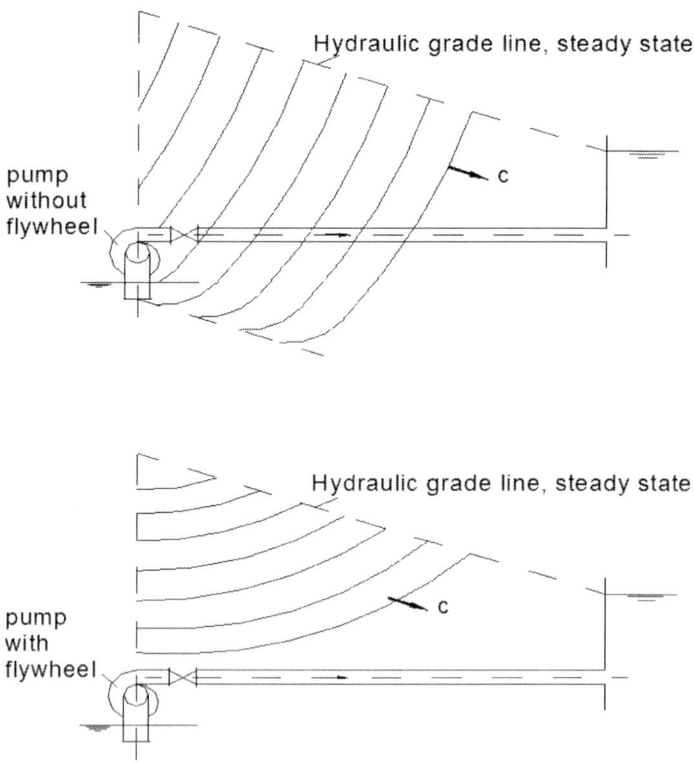

Figure 9. Effect of flywheel on transient pressure after power failure in the pumping station

A by-pass check valve is effective at sufficient suction pressure, which becomes available automatically in a booster station. Wavefront steepness is not affected until the by-pass check valve opens. A similar reasoning applies to the other pressure-limiting devices. Furthermore, the release of air pockets via air valves is an important source of inadmissible pressure shocks. Air release causes a velocity difference between the water columns on both sides of the air pocket. Upon release of the air pocket's last part, the velocity difference Δv must be balanced suddenly by creating a pressure shock of half the velocity difference (Figure 10). The magnitude of the pressure shock is computed by applying the Joukowsky law:

$$\Delta p = \pm \rho \cdot c \cdot \Delta v / 2 \qquad (5)$$

A large inflow capacity is generally positive to avoid vacuum conditions, but the outflow capacity of air valves must be designed with care.

3.4. Design Of Normal Procedures And Operational Controls

The following scenarios may be considered as part of the normal operating procedures (see also appendix C.2.2. in standard NEN 3650-1:2012):

1) Start of pumping station in a primed system.

2) Normal stop of single pump or pumping station.

3) Commissioning tests.

4) Priming operation or pump start in partially primed system.

5) Procedure to drain (part of) the system for maintenance purposes.

6) Normal, scheduled, valve closure.

7) Stop of one pumping station or valve station and scheduled start of another source.

8) Other manipulations that result in acceleration or deceleration of the flow.

9) Switch-over procedures.

10) Risk assessment of resonance phenomena due to control loops.

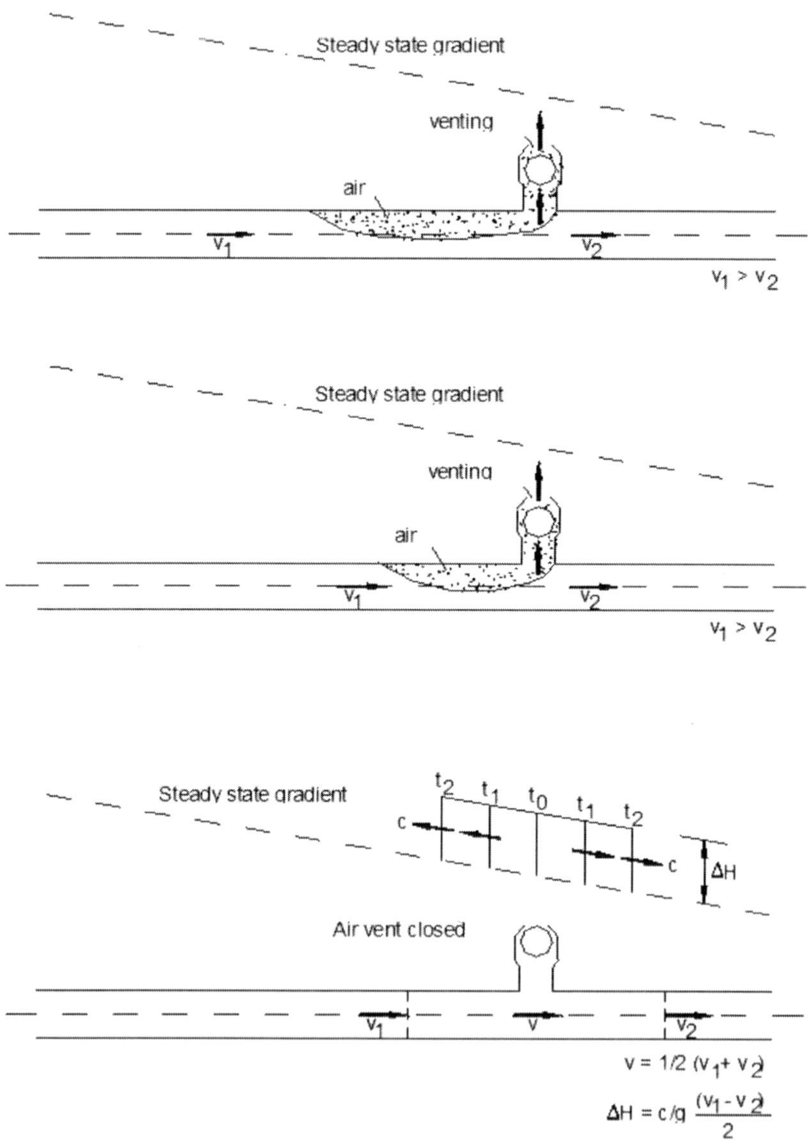

Figure 10. Pressure shock due to air valve slam.

Normal operating procedures should not trigger emergency controls. If this is the case, the control system or even the anti-surge devices may have to be modified. As a general rule for normal operations, discharge set-points in control systems tend to exaggerate transient events while pressure set-points automatically counteract the effect of transients. Two examples are given.

The first deals with a single pipeline used to fill a tank or supply reservoir. Suppose a downstream control valve is aiming for a certain discharge set-point to refill the tank or reservoir. If an upstream pump trip occurs, the control logic would lead to valve-opening in order to maintain the discharge set-point. This will lower the minimum pressures in the pipe system between the pumping station and the control valve. On the other hand, if the control valve aims for an upstream pressure set-point, the valve will immediately start closing as soon as the downsurge has arrived at the valve station, thereby counteracting the negative effect of the pump trip.

The second example is a distribution network in which four pumping stations need to maintain a certain network pressure. The pumping stations have independent power supply. Suppose that three pumping stations follow a demand prediction curve and the fourth pumping station is operating on a set-point for the network pressure. If a power failure occurs in one of the discharge-driven pumping stations, then the network pressure will drop initially. As a consequence the pump speed of the remaining two discharge-driven pumping stations will drop and the only pressure-driven pumping station will compensate temporarily not only the failing pumping station, but also the two other discharge-driven pumping stations. If all pumping stations would be pressure-driven pumping stations, then the failure of a single pumping station will cause all other pumping stations to increase their pump speed, so that the loss of one pumping stations is compensated by the three others.

The simulation of the normal operating procedures provides detailed knowledge on the dynamic behaviour of the WSS. This knowledge is useful during commissioning of the (modified) system. For example, a comparison of the simulated and measured pressure signals during commissioning may indicate whether the system is properly de-aerated.

It is emphasized that a simulation model is always a simplification of reality and simulation models should be used as a decision support tool, not as an exact predictor of reality. The design engineer of complex WSS must act like a devil's advocate in order to define scenarios that have a reasonable probability of occurrence and that may lead to extreme pressures or pressure gradients.

4. MODELLING OF WATER SUPPLY SYSTEMS FOR TRANSIENT ANALYSES

This section provides some guidelines on the modelling of a pipeline system with respect to pressure surge calculations.

It is recommended to model the top of the pipes in computer models, because the dynamic behaviour may change significantly at low pressures due to gas release or cavitation.

The modelling and input uncertainties raise the question of which model parameter values should be applied in a particular simulation. The simulation results may be too optimistic if the model parameters are selected more or less arbitrarily. The model parameters should be selected such that the relevant output variables get their extreme values; this is called a conservative modelling approach. The conservative choice of input parameters is only possible in simple supply systems without active triggers for control procedures. Table 4 lists the parameter choice in the conservative modelling approach.

If control systems are triggered to counteract the negative effect of critical scenarios (pump trip, emergency shut down), then the extreme pressures may occur at other combinations of input parameters than listed in Table 4. Therefore, a sensitivity analysis or optimisation routine is strongly recommended to determine extreme pressures in these kind of complex water supply systems.

5. CONCLUDING REMARKS

Since flow conditions inevitably change, pressure transient analysis is a fundamental part of WSS design and a careful analysis may contribute significantly to the reduction of water losses from these systems. It is shown that pressure transient analyses are indispensable in most stages of the life cycle of a water system. Section 2 shows that existing standards focus on a certain maximum allowable incidental pressure, but also emphasises that other evaluation criteria should be part of the surge analysis, including minimum pressures, component specific criteria and maximum allowable shock pressures. It is recommended that pressure shocks due to cavity collapse, air-release or undamped check valve closure should never exceed 50% of the design pressure. The main contributions of this paper, as compared to existing pressure transient design guidelines, include an overview of emergency scenarios and normal operating procedures to be considered, as well as the integrated design of control systems and anti-surge devices. These will lead to a safe, cost-effective, robust, energy-efficient and low-leaking water system.

Table 4. Overview of conservative modelling parameters for certain critical scenarios and output criteria.

Critical Scenario	Output Criterion	Model Parameters (conservative approach)
any operation (cavitation not allowed)	max. pressure and min. pressure	high wave speed or low wave speed, high vapour pressure
upstream valve closure or pump trip (cavitation allowed from process requirements)	max. pressure due to cavity implosions	high vapour pressure
upstream valve closure or pump trip	min. pressure	high friction and low suction level
downstream valve closure	max. pressure	high friction and high suction level
upstream valve closure or pump trip (surge tower present)	min. pressure and min. surge tower level	low friction and low suction level
downstream valve closure (surge tower or present)	max. pressure, max. surge tower level	low friction and high suction level
critical operation	criterion	model parameters (conservative approach)
upstream valve closure or pump trip (air vessel present)	min. air vessel level	low friction and low suction level and isothermal air behaviour

upstream valve closure or pump trip (air vessel present)	min. pressure (close to air vessel)	low friction and low suction level and adiabatic air behaviour
upstream valve closure or pump trip (air vessel present)	min. pressure (downstream part)	high friction and low suction level and adiabatic air behaviour
downstream valve closure (air vessel present)	max. air vessel level	low friction and high suction level and isothermal air behaviour
downstream valve closure (air vessel present)	max. pressure (close to air vessel)	low friction and high suction level and adiabatic air behaviour
downstream valve closure (air vessel present)	max. pressure (upstream part)	high friction and high suction level and adiabatic air behaviour
Single pump trip, while others run	max. rate of fluid deceleration	high friction and low suction level

REFERENCES

1. Boulos, P. F., B. W. Karney, et al. (2005). "Hydraulic transient guidelines for protecting water distribution systems." Journal / American Water Works Association 97(5): 111-124.

2. Jung, B. S. and B. W. Karney (2009). "Systematic surge protection for worst-case transient loadings in water distribution systems." Journal of Hydraulic Engineering 135(3): 218-223.

3. NEN (2012). Requirements for pipeline systems, Part 1 General. NEN, NEN. 3650-1:2012.

4. Pejovic, S. and A. P. Boldy (1992). "Guidelines to hydraulic transient analysis of pumping systems."

5. Pothof, I. W. M. (1999). Review of standards and groud-rules on transients and leak detection. Computing and Control for the Water Industry. Exeter, RSP Ltd, England.

6. Pothof, I. W. M. and G. McNulty (2001). Ground-rules proposal on pressure transients. Computing and Control for the Water Industry. Leicester, RSP Ltd, England.

7. Streeter, V. L. and E. B. Wylie (1993). Fluid transients in systems. New York, Prentice-Hall.

8. Thorley, A. R. D. (2004). Fluid Transients in Pipeline Systems. London, UK, Professional Engineering Publishing Ltd.

9. Tullis, J. P. (1989). Hydraulics of pipelines, pumps, valves, cavitation, transients. New York, John Wiley & Sons.

Chapter 2

Development of a Strategic Planning Model for a Municipal Water Supply Scheme Using System Dynamics

Adelere E. Adeniran[1*], Olufemi A. Bamiro[2]

[1]Civil & Environmental Engineering Department, University of Lagos, Lagos, Nigeria

[2]Mechanical Engineering Department, University of Ibadan, Ibadan, Nigeria

ABSTRACT

This paper reports the development of a system dynamics model for the strategic planning of a municipal water supply scheme. The model is capable of handling the critical variables that impact on the operations of a water supply scheme. The model was developed using the dynamo programming software to handle the process dynamics. Other supporting softwares—Fortran, Clipper 5, and Epiglue are used to handle data input, analysis and presentation of results in a user- friendly environment. The model was validated by applying it to the simulation of the University of Ibadan Water Supply Scheme for which extensive data of operations were collected over a period of ten years. The results obtained were found to be realistic and useful for the planning of plant operations. It is concluded that the model developed is sensitive to variable changes and has the capability of answering various operational "what-if" questions. As a strategic planning tool, the present model generates, for different scenarios, important operational information for the management of the water supply scheme. It is concluded that SD model is a useful tool to assist water managers and policy makers in making decisions and evolving strategic planning for water supply.

KEYWORDS

Municipal Water Supply, System Dynamics, Strategic Planning, Policy Decisions

1. INTRODUCTION

The city of Ibadan has been faced with water scarcity since the late 1980s. The scarcity of water in the city affected the University of Ibadan (UI) culminating in a fire incident that gutted a female undergraduate hostel. The management and operation of a Water Supply Scheme in an African metropolis is a complex and challenging problem that has cultural dimensions and demands a high level of strategic planning. A tool that can facilitate the performance of these functions is a step in the right direction [1] .

The management of the scarce water requires the development and operation of a model that can handle the complex dynamics that is involved [2] . In this paper, a System Dynamic (SD) model is used to capture the dynamics and the structure of the UI water supply system. The developed model was tested as a tool for making strategic planning decisions using simulations to replicate various scenarios. The objective of such a Strategic Planning Model for a Municipal Water Supply Scheme (SPMWSS) was to develop a strategic planning tool that would give the water managers, planners, policy and decision makers, flexibility in assembling and testing water management strategies for the university and similar settlements. A strategy may include structural measures to increase supply capacity as well as managerial options to manage water supply, reduce costs and to allocate water supply to various demand units at varying quantities and time [3] .

The simulation system is designed with the appreciation that there are several limitations on the processing and computing abilities of human managers and decision makers. A model for water supply management must necessarily focus attention on the flow of information in a complex water supply management system, and the use of such information in the decision making process. It should also focus attention on the role that simulation can play in gaining insights into current and future water management policies and strategies [4].

Three types of information can be used in the development of a strategic planning model: mental, written and numeric [5] [6] . In the development of the SPMWSS model, the mental information relied on the mental data base of the authors and those of their colleagues in the water unit of the University and it reliably captured the structure and policies governing management of the water supply scheme in UI. On the other hand, written data on water supply problem and modelling strategies abound. Several articles and conferences have tried to focus attention on paradigm shift and the need to apply modelling techniques to strategic planning in the water supply industry. The principles of SD are well suited for modeling and application to water resources and environmental problems [7] - [9] . SD is as an emerging tool with great potentials for improved understanding of environmental systems [10] . Numerical data used in building the model were collected and reviewed. For our problems, these included data from 1991 to 2003 kept by the UI water unit and unpublished reviews from consultants engaged at different times for improvement to the university's water supply.

2. CONCEPT, STRUCTURE AND DEVELOPMENT OF THE SPMWSS

The simulation model was formulated base on System Dynamics theory. System Dynamics, as an aspect of system theory, is a method for understanding the dynamic behavior of complex systems. The governing equations represented by finite difference expressions used for modeling different elements in a system are solved using standard numerical schemes [11] . System Dynamics modelling has attracted considerable attention over the past few decades. In a particular sense, System Dynamics is concerned with the use of models and modelling techniques to analyse complex systems and policy issues with a view of getting the right combination of scenarios to accomplish an efficient strategic planning for the system being modelled [3] .

In the SPMWSS model, DYNAMO, FORTRAN, CLIPPER 5 and EPIGLUE programming languages are combined to develop a user-friendly and menu-driven environment to drive the model. The simulation model consists of four major sectors.

The finance module includes details of the capital, operational and maintenance financial analysis. It is further sub-divided into (i) total cost in a 12-month cycle which can be used for budgeting and (ii) production and unit cost analysis which can be used for tariff setting mechanisms. The production module includes the model representation of population, per capita demand and supply and demand management options. The operation and maintenance module captures the pumping operations, availability, spare parts and maintenance schedules. It also includes store and inventory schedules. The distribution module lists the various water uses and demand nodes and their usage potentials. It incorporates the reservoirs and the pumping stations. Each module is a stand- alone but is interlinked with other module. Each of the modules also represents an aspect of water management from the point of view of sectoral interdependency (Figure 1).

The four modules are then housed in the SPMWSS model as interactive and interdependent components for the strategic planning of a Municipal Water Supply constituting the global system.

The concept of the SPMWSS model is as illustrated in Figure 2. The total model is segregated into three stages viz.: (i) Supporting Data Interface, (ii) Main SPMWSS Simulation Model and (iii) Report Generator Systems. The three-stage model concept is based on the identified interdependencies. In the SPMWSS, the DYNAMO language was adopted to model the various components and variables identified. The Equations are based on the mathematical philosophy of System Dynamics, the interconnectivity of the Water Supply System and the Syntax of DYNAMO modelling. The complete DYNAMO program listing (UIWATER.DYN) for the SPMWSS and the exogenous data file (UIWATER.ASC) is presented in [3] .

3. MODEL VALIDATION

The SPMWSS model is validated based on the rendition that the model must have the ability to replicate the real life system being modeled under the same parametric conditions. It is the validation of the basic underlying processes of the model (Design Validity). Selected endogenous variables, which are representative of the water supply system's behaviour, which have been collected from 1991 to 2003, are used to validate the model over a 12-month planning horizon each. Comparison of the model output with actual historical data over the same time period provides the required validation.

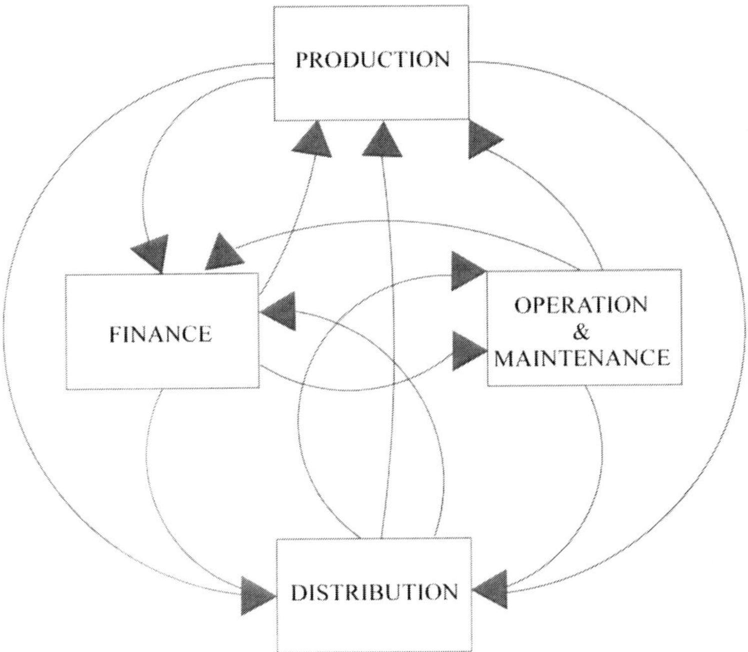

Figure 1. Interconnectivity of the major components of the SPMWSS model. Source: Adeniran (2007).

Figure 3 shows the comparison of the Actual Production for the years 1992 to 2002 under the real life operating conditions as available in the Water Works records as compared with the simulated outputs of the SPMWSS model for the same period. A comparison of simulated versus actual data shows that the variance is less than 5% and hence considered within reasonable limits.

Figure 4 compares the Actual Monthly production as available in the Water works records with the SPMWSS simulated Production under the same operating conditions for the year 2002. It is concluded from the two examples that the model replicates the system fairly well.

4. APPLICATION OF SPMWS: SCENARIO INVESTIGATIONS

Since the primary objective in building SPMWSS is the conduct of scenario experiments for strategic planning, considerable thought is given to the types of experimental features that must be built into SPMWSS to ensure effective and efficient scenario simulations can be carried out. The utilization of SPMWSS is oriented toward scenario experiments to production, finance and maintenance situations that will assist Management in short term and long-range and strategic planning of the water supply system. Scenario investigations allow the model users to ask "what-if" questions. The following strategic "what-if" questions were investigated using the model and the results presented.

Scenario 1: What if population should increase but per capita demand is constant?

Scenario 2: What if the plant efficiency should depreciate?

Scenario 3: What if there is a decline in raw water supply?

Scenario 4: What if the cost of energy (fuel and power) should increase?

Scenario 5: What if the plant efficiency should decrease?

Scenario 6: What if cost of raw water input from Eleyele intake (one of the raw water intake point), should increase?

Scenario 7: What if level of maintenance of the production plant should increase in order for production to remain constant?

Scenario 8: What if certain chemical dosage is applied, how much chemical?

Scenario 9: What if the production budget is required?

Scenario 1: Population Increase but Constant per Capita Demand

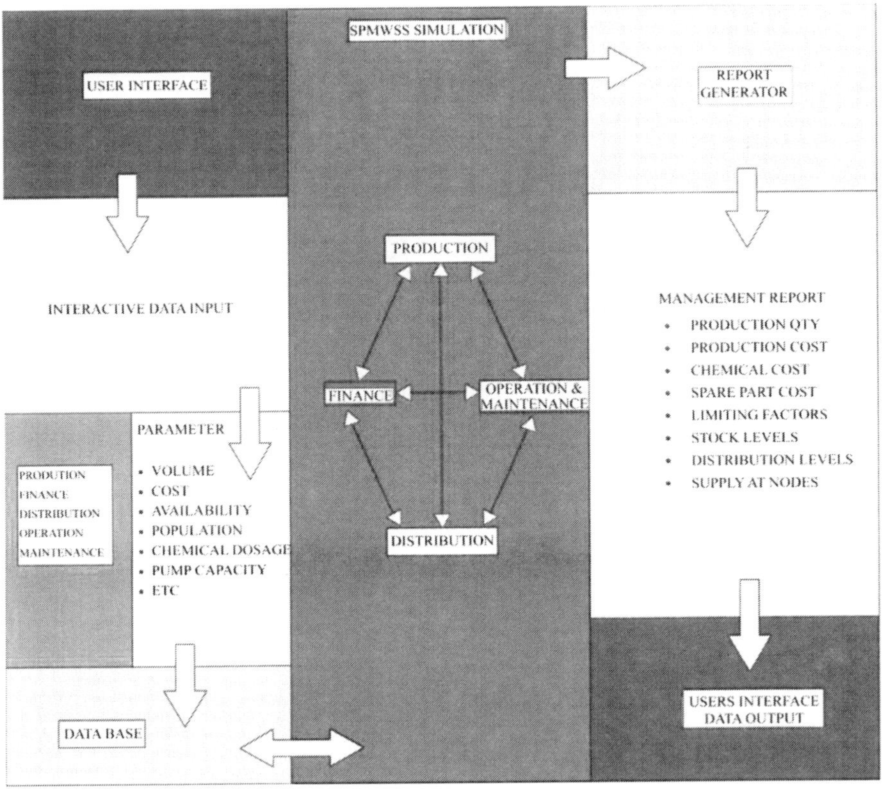

Figure 2. Concept of the SPMWSS model. Source: Adeniran (2007).

The simulation results of the above scenario are presented in Figure 5. The population is varied from 20,000 to 55,000 while the per capita water demand is kept at the WHO minimum of 110 litres. The cumulative annual productions are compared with the cumulative required production for each of the simulated situation.

It is observed from Figure 5 that the Water Supply Scheme is capable of meeting the per capital demand of 110 litres per person per day up to a population of about 35,000 after which it can no longer cope.

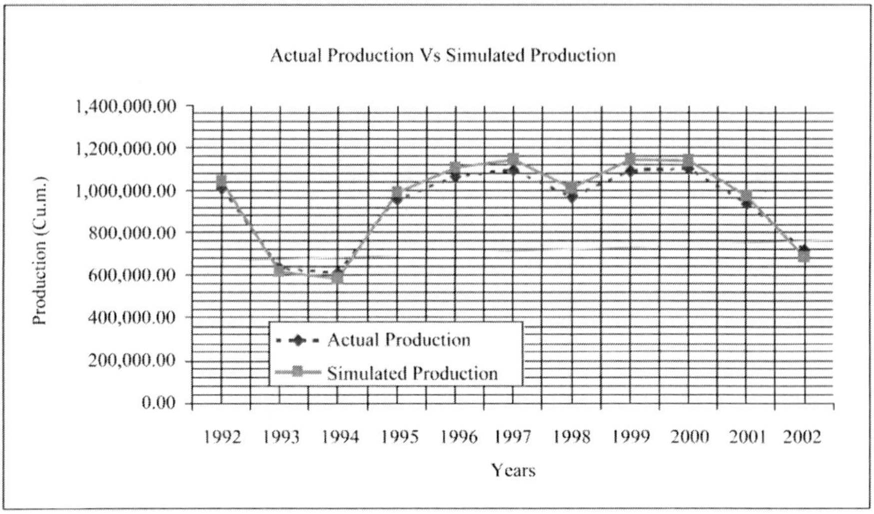

Figure 3. Chart of actual production vs. simulated production.

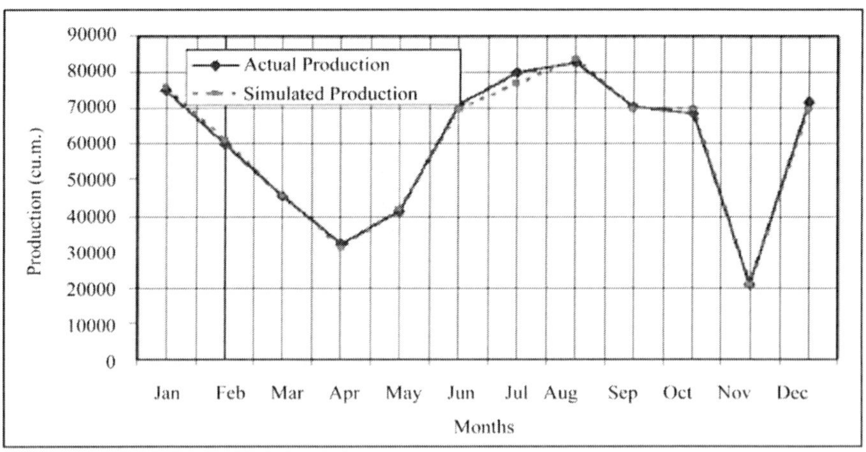

Figure 4. Year 2002 actual production vs. simulated production.

Scenario 2: Effect of Depreciating Plant Efficiency

One of the major factors militating against the ability of a water supply scheme to meet with the demand of the consumers is equipment depreciation and/or lack of maintenance resulting in decrease in Plant Efficiency. The effect of plant efficiency on water production is examined vis-à-vis the demand for water by the population. For this case, the population and per capita demand are kept constant while the plant efficiency is varied. The observations are presented in graphical form as Figure 6.

It is observed from Figure 6 that the demanded quantity of water is the same as the quantity of water produced when the efficiency of the plant is higher than 65%. In other words, the plant is capable of meeting the water demand of the population provided the efficiency is not below 70%. However at lower efficiency, the plant would not be able to cope with the demanded water by the population. It should be noted that for this scenario investigation, the population and the per capita demand are kept constant.

The strategic planning implication of this is that regular maintenance activities should be carried out to ensure that the efficiency of any the components of the system does not drop below 70%. This can be achieved by conducting regular reliability tests on the equipment.

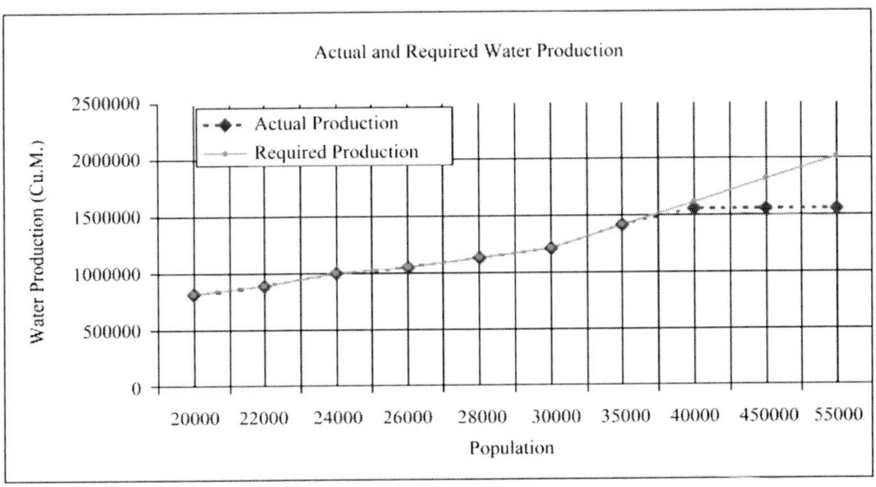

Figure 5. Actual production compared with water demand with varying population.

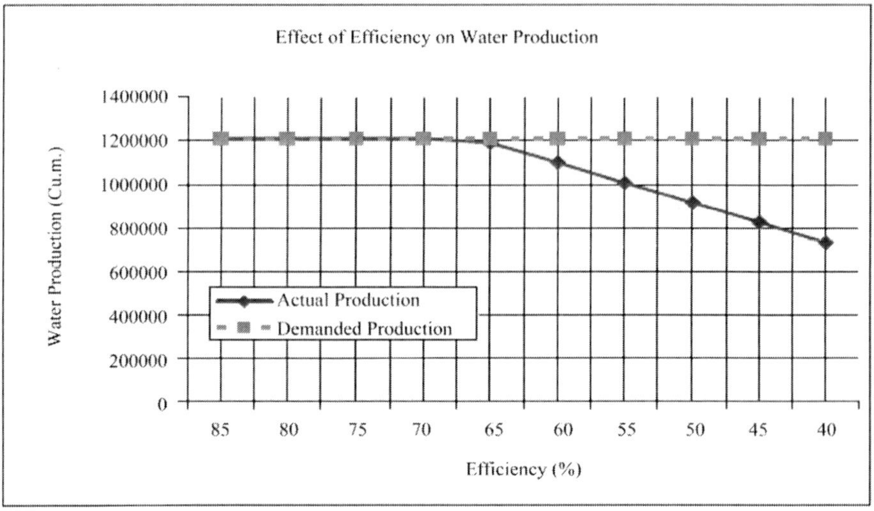

Figure 6. Effect of plant efficiency on water production.

Scenario 3: Declining Raw Water Supply

At present, the UI Water supply system modeled obtains raw water from both Eleyele and Oba Dam. While Eleyele is an external supply source, the Oba Dam is an internal supply source. A situation that can arise is for Eleyele source may not be available or the Oyo State Government who is the owner of the dam could come up with an exclusion policy and the University is stopped from the abstraction of raw water from Eleyele Dam. Presently 60% of the University's raw water is from Eleyele. In order to simulate the above condition, the scenario in which the raw water input of Eleyele dam is reduced gradually from 60% to 0% is simulated. This is done by reducing the hours of operation of the Eleyele pumps form 20 hours a day to zero hours a day. For this scenario, the population is assumed to be 30000 and per capita demand is 110 litres. The effect of this scenario on water production is as shown in Figure 7.

From Figure 7, the limiting factor becomes 1.0 when the raw water contribution from Eleyele becomes 12.5%. Any further reduction implies that the water demand of the population cannot be met. In fact, at 0% contribution from Eleyele, the limiting factor is 0.77. In other words, Oba Dam raw water source alone can cope with only 77% of the water demand of the population.

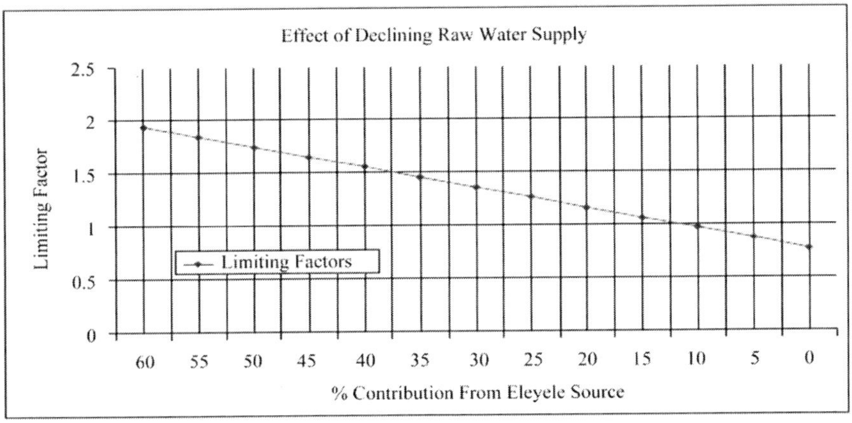

Figure 7. The effect of declining raw water supply on production.

The strategic planning implication is for Management to plan ahead and look for alternative source of raw water. For instance, the long proposed Ona river dam (see Figure 8) could be a feasible option to counteract the effect of declining raw water supply either from Eleyele or Oba dam.

Scenario 4: Increase in Cost of Fuel and Energy

Some of the major costs of production include the cost of Fuel and Energy (PHCN). From 1992 to 2005, the Nigerian Economy witnessed many fluctuations in the cost of energy and fuels. The effects of these changes are examined using the model. First, the effect of changes in the cost of fuel alone is examined, then the effect of the cost of energy (PHCN) alone is examined, and then the combined effects of these changes are examined together. The simulation results of these effects on the cost of Production and the Unit cost of production are presented in the following sections.

(a) Effect of Fuel Price on Unit Cost of Production

Table 1 shows the effect of the various changes in the petroleum product prices between 1992 and 2005. It is seen that the unit cost of production rose from N55.68 to N97.17.

(b) The Effect of Cost of Energy on Unit Cost of Production

Table 2 shows the effect of the various changes in the unit cost of (PHCN) energy N/kWh between 1992 an 2005. It is seen that the unit cost of production rose from N59.12 to N85.48. This is a single factor simulation while the other factors are kept constant.

(c) The Combined Effect of Changes in Cost of Fuel and Energy

Usually, the changes in a factor lead to changes in other factors in real life. Especially, whenever there is a change in the cost of petroleum products there is usually a correspondence change in the cost of energy and vice versa. Figure 9 shows the changes.

It is seen that the cost of production increases with increases in cost of fuel and energy. This result is expected, but with the SPMWSS model, it is now possible to quantify the effect of any slight changes in any of these important factors. It should be noted that these results are for a fixed population and fixed production quantity. The model is also capable of examining the effect of the economy of scale with varying production quantities [1] , [2] .

Scenario 5: Declining Plant Efficiency on Unit Cost of Production

One of the major factors in cost consideration is the efficiency of the production plant. The reduction in the plant efficiency does not only affect the production output it also affect the cost of production. In order to observe the effects of the efficiency on the cost of production, the efficiency is gradually reduced from the normal operating efficiency of 85% to 5%. The results of simulation are presented in the Figure 10.

It is seen from simulation results that the unit cost of production increases exponentially with linear declining of Plant Efficiency. It is obvious that it is not wise to operate the plant at low efficiency. The strategic plan implication of this is for adequate attention to be placed on maintenance such that efficiency is not allowed to drop. Any equipment that is operation below 70% efficiency should either be overhauled or replaced.

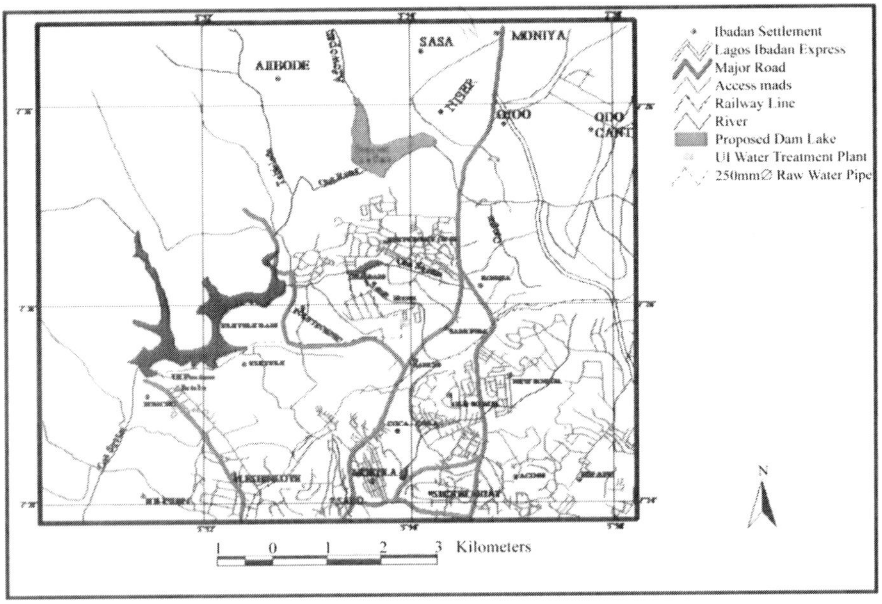

Figure 8. Map showing the locations of water supply reservoirs to university of Ibadan. Source: Works and Main- tenance Department, University of Ibadan.

Scenario 6: Cost of Raw Water Input from Eleyele Intake

Raw water is a major production input. At present, the University abstracts raw water from two sources, Eleyele Dam belonging to the Water Corporation of Oyo State (WCOS) and the Oba dam which belongs to the University. Payment of raw water to the WCOS has been very controversial from the inception of the University Scheme in 1990. The WCOS has insisted on payment for raw water and has increased the raw water tariff from N1.50, to N3.0 and now N6.0 per cu.m. In order to examine the effect of increase of raw water input from Eleyele, a scenario is created in which the cost of raw water from Eleyele is increased from N3.0 to N100.00. The results are presented in Figure 11.

Table 1. The effect of changes in fuel prices on cost of production.

PMS Price =N=	0.70	3.23	11.00	25.00	20.00	22.00	26.00	34.00	49.90	50.50
AGO Price =N=	0.50	2.50	7.50	15.00	12.50	22.00	27.50	42.00	52.00	65.00
Unit Cost =N=	55.68	57.09	60.80	66.65	64.66	70.10	73.63	82.53	89.99	97.17

Table 2. Effect of changes in unit cost of energy on unit cost of production.

Year	Changing PHCN Price =N=/kWh	Corresponding Unit Cost =N=/cu.m
1992	1.50	59.12
1993	3.00	62.89
1998	4.50	66.65
1999	6.00	70.43
2000	7.50	74.18
2001	7.50	74.18
2005	12.00	85.48

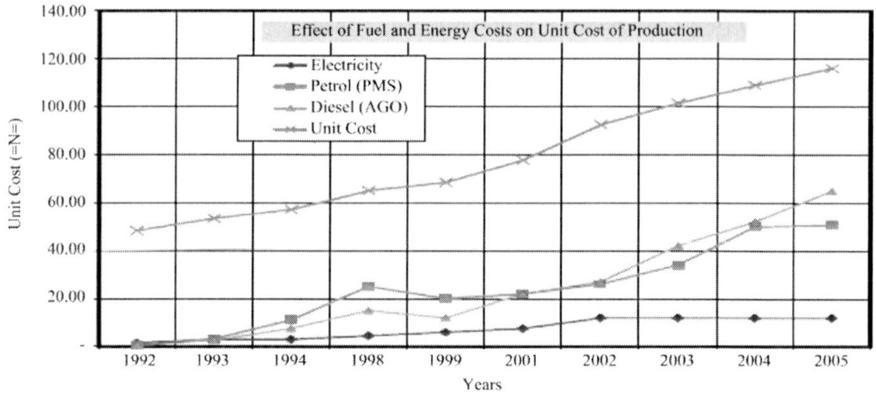

Figure 9. Effect of fuel and energy costs on unit cost of production.

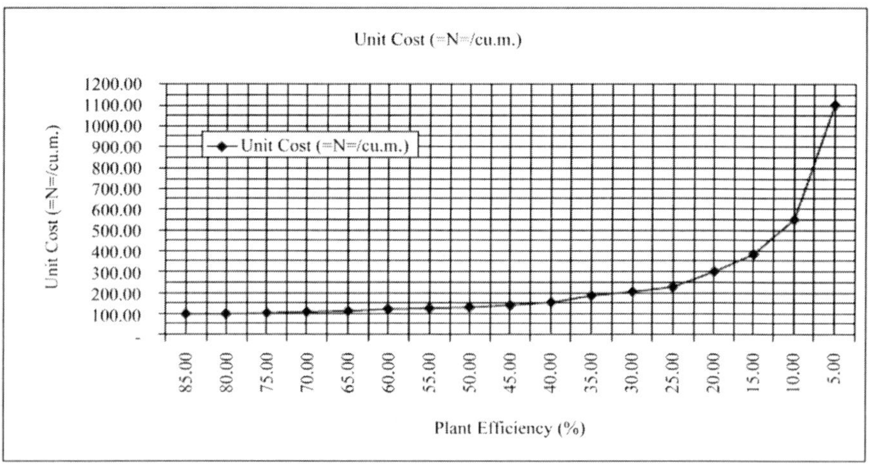

Figure 10. The effect of declining plant efficiency on unit cost of production.

It is seen that the production cost increases as the cost of raw water input from Eleyele increases. The Strategic Planning implication is evident from this graphic example: Suppose the WCOS decides to increase the cost of raw water from the present N6.00 per cubic meter to say =N= 100.00 per cubic meter, the unit cost of production will change from N92.11/m^3 to N234.53/m^3. This type of information can be used as argument against an arbitrary increase in the cost of raw water. It can also be used to justify the need for the University to look for its own internal sources of raw water.

Scenario 7: Increase in Level of Required Maintenance to Sustain Expected Production

The plant capacity is usually fixed relative to the level of maintenance input. If the plant is to be expected to have a constant production level, then the maintenance activities on the plant must be kept regular. Managers and operators are often requested to keep the plant running without corresponding supply of spare parts to ensure the optimum performance of the plant. Effective and cost effective maintenance also involve keeping reasonable level of staff for the running of the plant as well as evolving control policies as to the level of fuel consumptions and running cost expenses. Strategic planning in the area of maintenance, therefore, places primary emphasis on cost control with respect to the level of maintenance for constant production output.

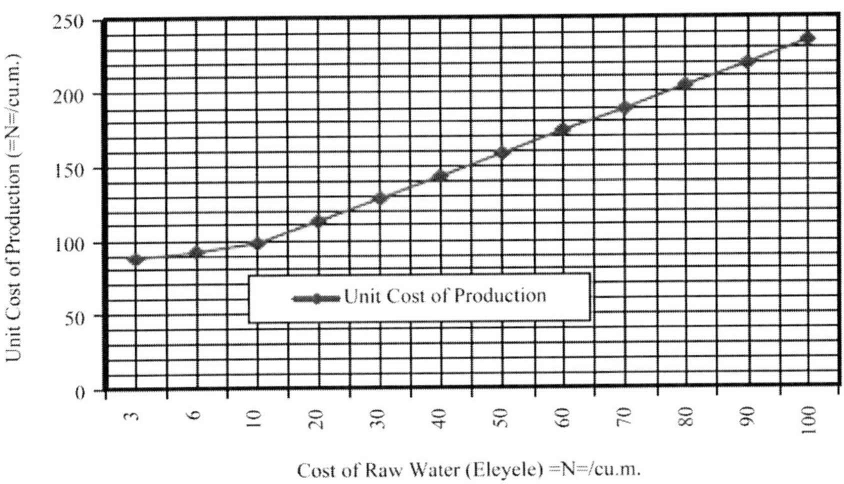

Figure 11. Effect of cost of raw water from Eleyele on unit cost of production.

The required level of maintenance of a system is a function of the maintenance input into the system. A new system requires a relatively low level of maintenance whereas a system that is tending towards unserviceability requires a very high level of maintenance. The aggregate of spare part usage, manpower attendance and down times are captured in terms of percentage. The required level of maintenance is varied from 10% to 100% while the production parameters are kept constant. The unit cost of production is observed for each of the simulated situation. The results are presented in Figure 12 and Table 3.

It is seen that the unit of production cost increases as the required level of maintenance increases. It is noted that when the level of maintenance is 0%, the unit cost is =N= 97.46 whereas when the level of maintenance required rises up to 100% i.e. requiring shut down maintenance, the unit cost of production is =N= 140.96.

It is also noted from Table 3 that if the level of required maintenance should increase from 0% to only 10%, the University has to look for an additional sum of =N= 5,000,000.00 in order to maintain production. The situation becomes worse with higher level of required maintenance. If the level of required maintenance is allowed to reach 100%, the university must look for as much as =N= 52,535,000.00 for turnaround maintenance. The strategic planning implication is that management must evolve and sustain a maintenance policy that ensures routine maintenance in order to limit the production cost. Under no condition must equipment be allowed to depreciate to the level of breakdown maintenance.

Scenario 8: Total Chemical Usage Given Chemical Dosage

As an Expert System, the SPMWSS can be used as an oracle to tell the total quantities of chemical expected to be used if the level of chemical dosages at each period of the planning horizon is known. The model can thus be used to monitor chemical usage of the treatment plant and thus prevent and or detect theft. In this example, Alum alone is used. However, other chemicals can be equally monitored. The dosage of Alum is varied from 80 ppm to 180 ppm based on the turbidity profile of the raw water which varies from season to season. The best Alum dosage is about 80 ppm while the worse is about 180 ppm at the peak of the raining season. Table 4 shows the expected Total Annual Alum usage at a given dosage. This information can be used to intelligently figure out the quantity of Alum that Management should purchase if the average dosage in a year is known.

Scenario 9: Expert System for Production Budget

One of the attributes of the SPMWSS model is its budgeting capability. In a public establishment, budgets are often required within a short time frame. Such budgets are usually prepared using the rule of the thumb. SPMWSS model is an Expert System for preparing production budget because all the financial components for preparing a production budget have been built into the model. In actual sense, the total cost under certain conditions can be given as a model output as shown inTable 5. It is seen from this table that cumulative costs of each of the vital production item for the simulated condition are presented. Thus it is known that about =N= 6,161,200 is required for Chemicals while about =N= 511,000.00 is required for spare parts and a total of =N= 117,716,000 is the total money required for the water sector during the planning year.

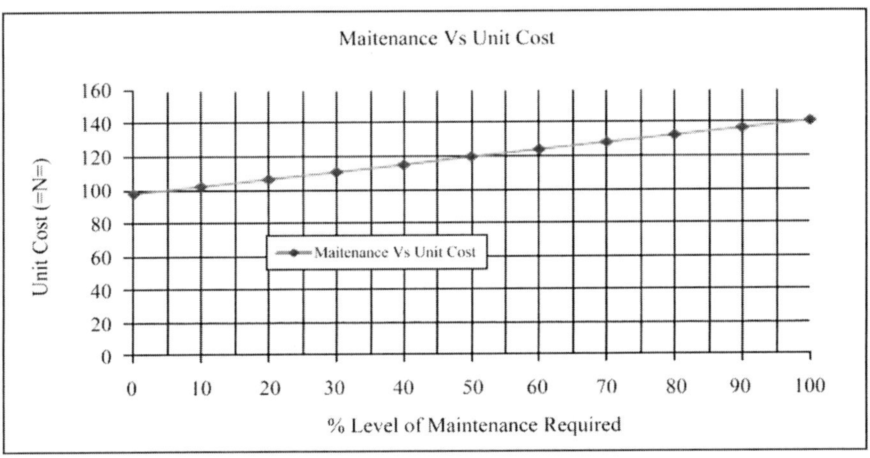

Figure 12. Effect of increased level of maintenance on unit cost of production.

Table 3. Effect of increased level of maintenance on cost of production.

% Level of Required Maintenance	Total Production Cost =N=:k	Unit Cost of Production =N=:k
0	117,716,000.00	97.46
10	122,977,000.00	101.82
20	128,218,000.00	106.16
30	133,468,000.00	110.51
40	138,711,000.00	114.85
50	143,976,000.00	119.21
60	149,242,000.00	123.57
70	154,486,000.00	127.91
80	159,727,000.00	132.25
90	164,979,000.00	136.59
100	170,251,000.00	140.96

Table 4. Total alum usage given dosage level.

Alum Dosage (ppm)	80	90	100	110	120	130	140	150	160	170	180
Alum Usage (Kg)	96,627	108,702	120,780	132,840	144,960	157,020	169,080	181,121	193,260	205,320	217,380

5. CONCLUSION

In this paper, the development of a computer based strategic planning model for a water supply scheme has been reported. The model is developed using the DYNAMO modeling software. However, in order to create a user- friendly environment, other application programs such as FORTRAN, CLIPPER 5 DBASE, EXCEL and EPIGLUE have also been used in an integrated manner. The model was validated using existing data that spanned ten (10) years from the University of Ibadan Water Supply Scheme. The model was then used to carry out scenario experimentations for strategic planning. The results presented indicate that an SPMWSS model has been successfully developed to handle the operation of the University of Ibadan Water Supply Scheme. The potential of greater understanding of a system through the system dynamics approach, which is still in its infancy, is shown to be not only feasible but also exciting. The model is found useful to managers and planners for its strategic planning capability. The SPMWSS model integrates the production and financial elements; in addition to achieving true total cost perspective, this particular capability introduces the potential flow paths from production to consumption. The SPMWSS provides a mechanism wherein the modeled system can be examined under scenario experiments. In addition, many other aspects of system management as they relate to water supply can be tested in an effort to evaluate existing policies and principles.

Table 5. Cumulative costs of various production item.

PERIOD	TOTAL EXP.	CHEN. COST	SPARE PARTS	INCIDENTALS	SALARY	POWER COST	TRANSPORT
₦	₦	₦	₦	₦	₦	₦	₦
1	9,723,000.00	454,400.00	5000.00	5000.00	940,000.00	3,081,000.00	4,067,000.00
2	18,901,000.00	879,500.00	16,000.00	10,000.00	1,880,000.00	5,963,000.00	7,872,000.00
3	28,645,000.00	1,333,900.00	42,000.00	15,000.00	2,820,000.00	9,044,000.00	11,939,000.00
4	38,120,000.00	1,773,800.00	74,000.00	20,000.00	3,760,000.00	12,026,000.00	1,587,500.00
5	47,879,000.00	2,264,000.00	79,000.00	25,000.00	4,700,000.00	15,107,000.00	19,942,000.00
6	57,396,000.00	2,773,200.00	84,000.00	30,000.00	5,640,000.00	18,089,000.00	23,878,000.00
7	68,156,000.00	3,370,800.00	260,000.00	35,000.00	6,730,000.00	21,125,000.00	28,559,000.00
8	7,891,600.00	3,968,400.00	436,000.00	40,000.00	7,820,000.00	24,161,000.00	33,240,000.00
9	88,497,000.00	4,546,900.00	436,000.00	45,000.00	8,760,000.00	27,143,000.00	37,176,000.00
10	98,370,000.00	5,144,500.00	448,000.00	50,000.00	9,700,000.00	30,224,000.00	41,243,000.00
11	107,930,000.00	5,671,000.00	479,000.00	55,000.00	10,640,000.00	33,206,000.00	45,179,000.00
12	117,116,000.00	6,161,200.00	511,000.00	60,000.00	11,580,000.00	36,287,000.00	49,246,000.00

REFERENCES

1. Adeniran, A.E. and Bamiro, O.A. (2010) A System Dynamics Strategic Planning Model for a Municipal Water Supply Scheme. Proceedings of the International Conference of the System Dynamics Society, Seoul, 25-29 July 2010.

2. Adeniran, A.E. (2014) Application of System Dynamics Model in the Determination of the Unit Cost of Production of Drinking Water. International Journal of Water Resources and Environmental Engineering, 6, 183-192.

3. Adeniran, A.E. (2007) Development of a System Dynamics Strategic Planning Model for a Municipal Water Supply System. Ph.D. Thesis, University of Ibadan, Ibadan.

4. World Bank (1993) A Strategy for Managing Water in the Middle East and North Africa, IBRD. World Bank, New York.

5. Forester, J.W. (1961) Industrial Dynamics. MIT Press, Cambridge.

6. Forrester, J.W. (1992) Policies, Decisions and Information Sources for Modeling. European Journal of Operation Research, 59, 42-63.

7. Fletcher, E.J. (1998) The Use of System Dynamics as a Decision Support Tool for the Management of Surface Water Resources. First International Conference on New Information Technologies for Decision Making in Civil Engineering, Montreal, 11-13 October 1998, 909-920.

8. Ford, A. (1999) Modeling the Environment—An Introduction to System Dynamics Modeling of Environmental Systems. Island Press, Washington DC.

9. Deaton, M.L. and Winebrake, J.I. (1999) Dynamic Modeling of Environmental Systems. Springer-Verlag, New York.

10. Huang, G.H. and Chang, N.B. (2003) Perspectives of Environmental Informatics and Systems Analysis. Journal of Environmental Informatics, 1, 1-6.

11. Bamiro, A.O. (1995) Electronic Oracle—Seeing through Computer Models. Post Graduate School Interdisciplinary Discourse, University of Ibadan, Ibadan.

Chapter 3

Model Based Sustainable Management of Regional Water Supply Systems

Thomas Bernard[1], Oliver Krol[1], Thomas Rauschenbach[2] and Divas Karimanzira[2]

[1] Business Department MRD, Fraunhofer Institute of Optronics, System Technologies and Image Exploitation IOSB, Karlsruhe, Germany

[2] Fraunhofer Application Center System Technology AST, Fraunhofer Institute of Optronics, System Technologies and Image Exploitation IOSB, Ilmenau, Germany

1. INTRODUCTION

The sustainable management of the water resources and a safe supply of drinking water will play a key role for the development of the human prosperity in the following decades. The fast growth of many cities puts a large pressure on the local water resources, especially in regions with arid or semi-arid climate. A research project at Fraunhofer IOSB and AST has aimed to investigate ways for economic and sustainable use of the available water resources in the region of the capital of China, Beijing [1].

A main issue of the project is to develop components for a model-based decision support system (DSS), which will assist the local water authority in management, maintenance and extension of the water supply system at hand [2]. This paper deals with the derivation of suitable management strategies for a mid (till long) term horizon based on assumptions for future environmental and socio-economic conditions, which are provided by other modules of the DSS. The general structure of the proposed optimal control DSS is shown in Fig. 1.

An overview about several DSS concepts and implementations is provided in [3, 4]. A challenge of the given problem is the large area of the water supply system. The water management has to consider the total water resources of five river basins with an area of 16,800 km^2 as well as large groundwater storage in the plain with an area of 6,300 km^2. The main portion of the annual precipitation (85%) in this semi-arid region is falling from June to September leading to a highly uneven distribution throughout the year. The formerly abundant groundwater resources have been overexploited over the last decades resulting in a strong decline of the groundwater head (up to 40 m). Five reservoirs are important for the management of the surface water in the considered area, where the two largest account for about 90% of the total storage capacity of roughly 9×10^9 m^3. The water is distributed to the customers using rivers and artificial transport ways (channels, pipes) of a total length of about 400 km.

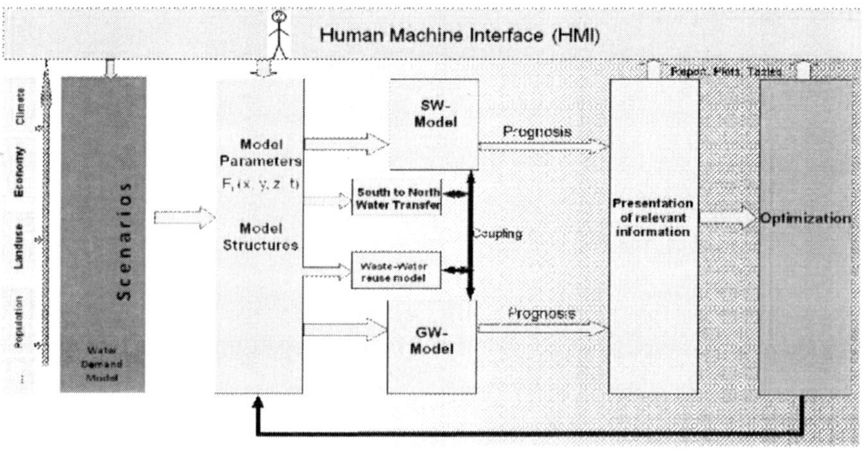

Figure 1. Structure of the proposed decision support system (DSS) for the region of Beijing

A common approach for policy generation in this field of application is to use mathematical programming techniques based on a dynamic model of the essential elements of the water allocation and distribution system [3].

The great impact of the groundwater storage for the supply system at hand requires a more detailed description of the groundwater flow dynamics compared to other known DSS implementations. Therefore a 3D Finite-Element model of the plain region has been developed. However, a direct integration of this 3D model into an optimal control framework is not possible due to its computational costs. A trajectory based model reduction scheme is proposed, which guarantees a very fast response of the DSS in combination with a specially tailored non-linear programming algorithm.

The chapter is organized as follows: In section 2 the essential models (surface and groundwater models) and the model reduction approach are described. While the formulation of the optimal control problem is subject of section 3, the numerical solution of the large scale structured non-linear programming problem is described in section 4. First results of the optimal water management approach are presented in section 5.

2. WATER ALLOCATION MODEL

The water allocation model can be divided into the surface water model and the groundwater model. The parts of the water allocation model are described in the sequel.

2.1. Surface Water Model

The surface water model has to comprise all important elements for the allocation, storage and distribution of water within the considered region. The intended field of application of the decision support system under development embraces management and upgrading strategies for the mid and long term range. This implies a significant simplification for the process model, because the retention time along the different transport elements (river reaches, channel or pipelines) is less than the desired minimum time step for decision of one day. Therefore, a simple static approach for flow processes is sufficient and the use of sophisticated models for the dynamics of wave propagation (like e. g. Saint-Venant-Equations) with respect to control decisions is avoided. In this case, the flow characteristics are represented by simple lag elements of the first order combined with dead-time elements. Given y and $u$as an output and input, respectively. The following mathematical relationship indicates a lag element of the first order.

$$T_1 \cdot \frac{dy}{dt} + y = u \tag{1}$$

where T_1 is the only parameter and indicates the time lag; t indicates the time. The following relationship represents the dead-time element:

$$y = u(t - T_t) \tag{2}$$

In this case the parameter is the dead-time *Tt*, which is thus the measure of the time taken for water to flow in a conduit over a known distance.

The surface water system is described as a directed graph. The edges characterize the transport elements and introduce only a variable for the discharge. The nodes represent reservoirs, lakes, points of water supply or extraction and simple junction points. Every node constitutes a balance equation involving the edges linked with and possibly the storage volume. The sole nonlinearity results from modeling the evaporation from the water surface of the storages (volume-area-curve), which is described by a piecewise polynomial approach. At time step *k*, the volume of a reservoir node evolves as follows:

$$V_j^{k+1} = V_j^k + \Delta t^k \left(\sum_{i \in E(j)} Q_i^k + A_{O,j}^k q_{evpot,j}^k - q_{seep,j}^k + A_{O\max} q_{prec,j}^k \right) , \tag{3}$$

where $A_{O,j}^k = f(V_j^k)$, the storage volume is denoted by *V*, the discharge into and from the storage is denoted by *Q* and discrete time step is denoted by *Δt*. The total evaporation from the reservoir depends on the water surface *A0* and the potential evaporation *qevpot.* q*seep* denotes the seepage from the reservoir to the groundwater and *qprec* specifies the precipitation.

Channels with a very low slope are modeled as water storage. The level dependent upper bound for the channel outflow is derived from a steady state level-flow relation like e.g. Chezy-Manning friction formula and is directly added as constraint to the optimization problem.

The Structure of the Beijing water supply system is shown in Fig. 2. Firstly, there are the four main reservoirs Miyun, Huairou, Baihebao and Guanting. Further sources are groundwater storages and water transfers. Secondly, there are the water transportation systems such as channels and rivers. Miyun reservoir and Huairou reservoir are connected to Beijing by the Miyun-Beijing water diversion. In the simulation model the arrows describe hydraulic behavior of water flow. Baihebao and Guanting reservoir are connected by tunnel and river Guishui. From Guanting water runs into the Yongding river water diversion system to Beijing. Existing retention areas for flood control are also considered in the simulation model.

The surface water from channels and rivers is delivered to the customers in different ways, directly or through the surface waterworks. Groundwater is distributed to the customers through ground waterworks as well as motor-pumped wells. Therefore, the waterworks build the third part of the Beijing water supply system. The last category is made up of all the customer groups (agriculture, industry, households and environment). To complete up the cycle, catchments area models are integrated in the system to take into account precipitation and evapotranspiration.

The surface water simulation model has been implemented in Matlab/Simulink using the toolbox "WaterLib" [5] and contains the most important elements of the drinking water supply system of Beijing. The simulation model was developed to reach a sufficient accuracy as well as a high simulation speed. A one-year simulation is carried out in a simulation time of several seconds. This is a very important condition for using the model in the decision support system.

2.2. Groundwater Water Model

The most important water resource in the considered area is groundwater that is modeled by a dynamic spatially distributed finite element groundwater model. The governing equation for groundwater flow is Darcy's law [6] describing slow streams through unconfined aquifers. Combining Darcy's law with mass conservation yields the partial differential equation (4) which is a diffusion equation.

$$S_0 \dot{h} - \nabla \cdot (k_f \nabla h) = Q_{rech} - Q_{\exp l} \qquad (4)$$

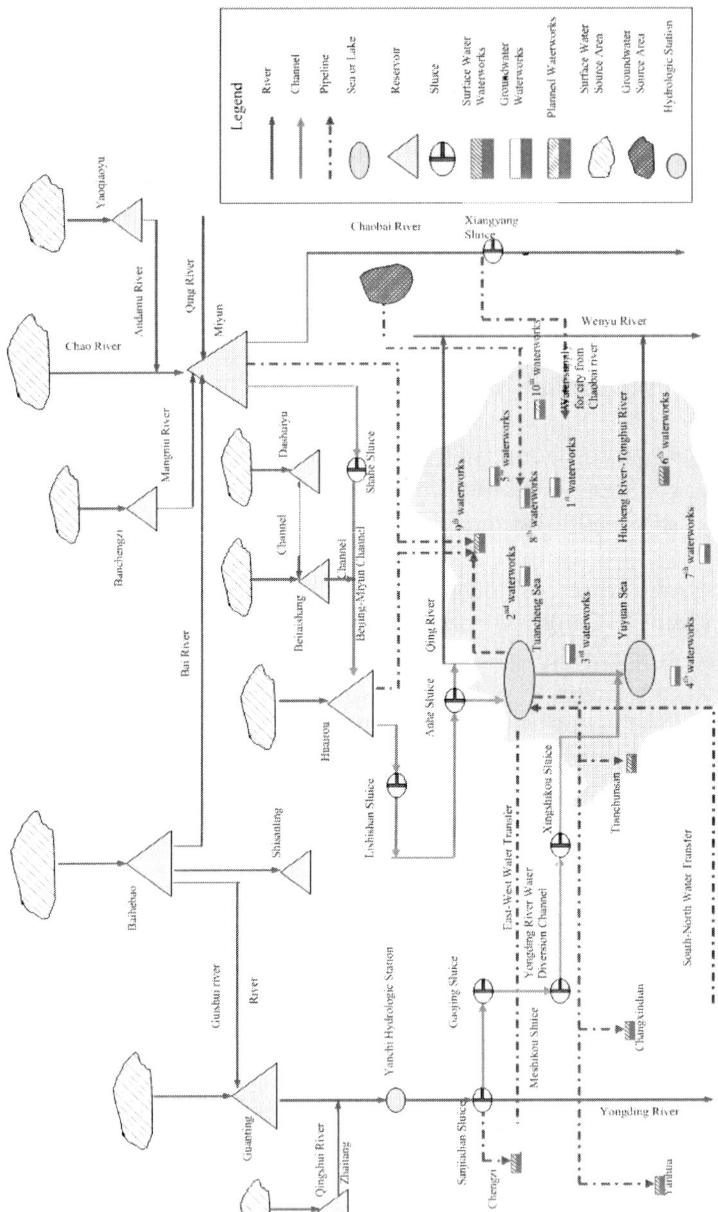

Figure 2. Structure of the Beijing water supply system

In (4) denotes *h* the hydraulic head (which corresponds to the groundwater level) and *kf* the hydraulic conductivity that governs the hydrogeological properties of the soil. S_0 denotes the specific storage coefficient. The terms on the right hand side of (4) summarize all sources and sinks that coincide with the time dependent groundwater exploitation due to industry, households and agriculture (Q_{expl}) and recharge e. g. due to precipitation and irrigation (Q_{rech}) in Ω.

The partial differential equation (4) is an initial-boundary value problem which has to be solved numerically for *h* in the 3 dimensional model domain Ω. The groundwater model has been implemented using FEFLOW, which is a Finite Element (FEM) software specialized on subsurface flow [7]. The initial condition is h (Ω, t_0) (groundwater surface) at the initial time t0. The inflow/outflow is described by Dirichlet boundary conditions, i.e. h (∂Ω) at the boundary ∂Ω and by well boundary conditions, that define a particular volume rate into or out of Ω. The advantage of the latter one is that they are scalable. The 3D FEM model consists of more than 150,000 nodes, distributed on 25 layers (cf. Fig. 3). Huge computational costs result from this high resolution. The simulation of 5 years needs ~15 Minutes on an Intel Core 2 Duo CPU (2.5 GHz). Hence it is very time consuming to calculate optimal water allocation strategies with the 3D FEM groundwater model. This is the motivation for model reduction (see subsection 2.3).

The main task with respect to the groundwater model is the parameterization of the large-scaled model covering an area of 6,300 km^2. On the one hand the time independent soil parameters k_f, S_0 have to be estimated and generalized for the whole domain Ω by a (small) set of measured values. On the other hand the source / sink terms Q_{expl}, Q_{rech} have to be calculated time dependent. For these calculations time dependent maps of precipitation and water demand are needed. The water demand is splitted into the three user groups households, industry and agriculture (see [8] for details). This parameterization issue is supported by powerful geographical information systems (GIS).

2.2.1. Hydrogeological Conditions And Derivation Of Hydrogeological Parameters

The groundwater model area is located at the northern part of the North China Plain (NCP), which is the largest alluvial plain of eastern Asia. The NCP is a basin with quaternary aged surficial deposits (loess, sand, gravel and boulder, silt and clay). According to the hydrogeological profiles of Beijing the quaternary system in this region is fairly complicated. A great variety of different sedimentary facies exists with different thicknesses ranging from several tens of meters around the piedmont area to 150 - 350 meters in the northern central part of the NCP [9]. Groundwater is exploited in the layers of quaternary deposits, i.e. in the loose stratum/porous aquifers with high to very high water storage capacities. From the Taihang Mountains in the west to east there are two main geomorphological units in the model area: the piedmont plain below the mountain escarpments and the flood plain. In the piedmont plain the aquifers structure is coarse and becomes finer from west to east. In the flood plain the structure of aquifer is fine with silt sand, clay and silt interlay and in areas of ancient rivers and paleochannels the aquifer is composed mainly of gravels and coarse sands with good permeability. Therefore the distribution of groundwater in the Beijing region is inhomogeneous. Regions of high abundance and high yielding porous groundwater aquifers are the piedmont plains and the northeastern districts of Miyun, Huairou and Shunyi whereas less yielding aquifers are found in the Yangqing and Tong districts. In the transition zone from the Taihang Mountains to the NCP the quarternary sediments with low thickness of e. g. some tens of meters are lying on the older rock formations of the regions.

In the mountainous districts unstable groundwater distributions were assumed in dependence on the form of the rocks with geological discontinuities (fractures, joints, dissolution features) and the groundwater flow. In the transition area from the Taihang and Yanshan Mountains to the NCP stratigraphic sequences of various ages ranging from archaean metamorphic rocks to quaternary are documented in the geological and hydrogeological maps. A detailed description of the geological and hydrogeological conditions can be found in [10].

On the base of a conceptual geological model and a structured horizontal (2D) groundwater model, a horizontal and vertical structured 3D - groundwater model was developed describing the saturated zone till approx. 200 m depth below ground surface (bgs.) in the area of the quaternary sediments of the NCP. In addition the borehole data from approx. 125 drillings situated in the model area were used in the groundwater model. Although a quite homogeneous distribution of the boreholes was given, one measurement point represents an area of about 50 km^2 which is only a rare database for modelling subsurface conditions.

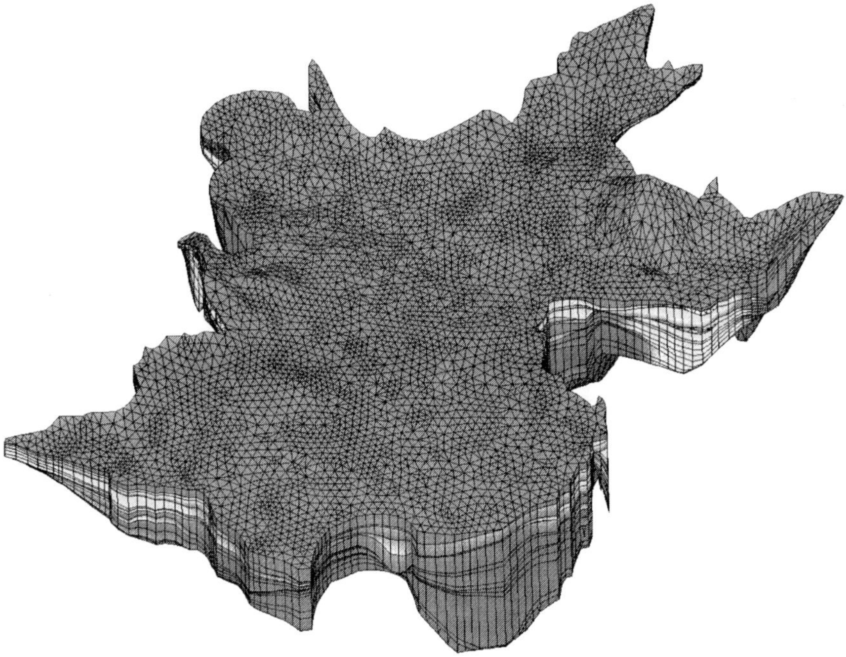

Figure 3. Mesh of the 3D Finite Element groundwater model of the region of Beijing

There can be found strong variations in structure and thickness of the loose stratum sediments in the model area. The evaluation of all data (borehole data, geological and hydrogeological maps and profiles, ground water levels from observation wells, literature etc.) shows that the large number of the water bearing layers can be summarised in up to three essential ground water aquifers according to present knowledge on regional level. These aquifer systems are from top to down:

- Aquifer I: Shallow aquifer in approx. 5 m to 30 m depth bgs.
- Aquifer II: Primary aquifer, till approx. 120 m depth bgs.
- Aquifer III in the depth area of approx. 120/140 m to 200/260 m bgs

The aquifers are separated by less permeable layers or aquitards, above all fine sands, silts and clays. It can be assumed that the three essential aquifers are not completely independent from each other, i.e. a groundwater exchange takes place between them in a certain range. Where low permeable layers or aquitards are absent or have a low thickness two aquifers can form a hydraulic unity as in the area of piedmont plains. Thus in the piedmont plains only one porous aquifer between the unsaturated loess top set layers and the bedrock was assumed. In regions, in which the separating layers have bigger thickness and larger extension, local confined aquifers can appear. Because of morphology and evolution processes perched aquifers can appear within the loess deposits. All these local effects are summarised in the above mentioned three essential aquifers.

The piedmont areas of the Taihang Mountains and the Yanshan Mountains along the western boundary and the northern/northeastern boundary of the area are the areas where groundwater inflow into the plains contributes to the groundwater recharge of the confined and unconfined aquifers. Because of the multi-layered geological structure of the loose stratum in the model area, consisting of loess, alluvial loess, sand-gravel-cobble-boulder sediments with more or less mighty clay and silt inclusions, the details of hydrogeological conditions are complicated. Therefore the used spatial distribution of the hydraulic conductivity and specific yield (both important for good model results) are adequate phenomenological descriptions of mean values and not a detailed representation of local conditions in reality.

The aquifer characteristics were determined mainly on the base of the interpretation and evaluation of the above mentioned borehole data. The borehole data represent the geological layers (boring logs) at single points. They show high variations from one point to another. In particular the hydrogeological parameters k_f and S_0 were deduced from these borehole data due to the following procedure:

1) In a first step the local single point data have to be transformed in values representative for an associated area (meso-scale values) [11]. For this step the information and data from thematic maps (e.g. Beijing hydrogeological maps lithology map, water abundance map, etc.) were included. The validity range of these meso-scale values could be specified with the informations from a water abundance map representing the water yield and water storage capability. These data resulted primarily on measurements in water exploitation wells. With this approach meso-scale values (for k_f -and S_0) and their spatial distribution could be determined.

2) In a second step the discrete meso-scale values were interpolated within the model area

3) In a third step the absolute values of the smoothed spatial distribution of k_f and S_0 were adapted to fulfil the water budget of the model area.

On the basis of the derived values a fine tuning was realized in order to get minimal differences between calculated and measured groundwater surface map.

The k_f -values of the aquifers range from $2.0x10^{-3}$m/s in the region of the piedmont plains and the alluvial fan plains to $0.1x10^{-4}$m/s in the flood plains. The loose stratum depositions can be classified according to German Institute for Standardization Guideline 18130 as permeable to very permeable.

In [12] a mean storativity in the range of 0.08 and 0.18 has been estimated. Due to hydrogeological investigations of borehole data a regionalisation of the hydrogeological parameters could be performed, yielding a mean storativity of 0.13. The values of S_0 changes from 0.19 (piedmont plains) to 0.03 (flood plains). This spatial distribution of k_f - and S_0-values was also a basis for the regionalisation of the inflow boundary conditions.

Inflow conditions

The inflow from the north and west into the model area is governed by the transition zone between the mountain terrain and the plain where a high conductivity can be assumed. Here the inflow consists of the surface water run off from the mountains that depends on the precipitation rate. But due to investigations even after a number of dry years the total inflow did not disappear, but decreased from a long term mean value of about $0.7x10^9m^3/a$ to $0.4x10^9$ m^3/a could be observed by the Chinese partners. This amount of water could not only result from rainfall runoff from the mountainous domain. Therefore it was assumed a split of the horizontal groundwater recharge from groundwater inflow into the model area (see Fig. 4). One part is a rain-dependent contribution which in 'normal years', with a mean precipitation rate of about 590 mm/a is in the range of $0.4x10^9$ m^3/a. This value is scaled in dependency of the mean precipitation rate of the current year. In a dry year with a precipitation rate of 380 mm/a, for instance, the rain dependent inflow to $0.26x10^9$ m^3/a. For this contribution the imagination is that a part of the precipitation of the mountain slopes infiltrate on the surface and percolate down to the bedrock basis. On the relatively impermeable bedrock the water flows as shallow groundwater aquifer in the loose stratum with low thickness into the model area. The loose stratum depositions in the piedmont plains consist of boulder, gravels and sand with inclusions of local loess and loessloam lenses. These loose stratum depositions are well permeable and this inflow contribution from the mountains is asssumed to be directly dependent on the precipitation.

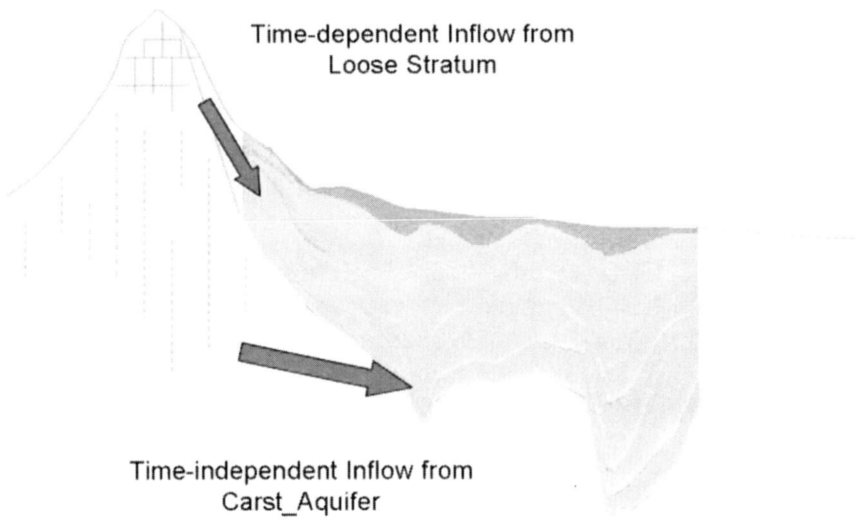

Figure 4. The groundwater inflow regime into the NCP

The second contribution to the horizontal groundwater recharge from the mountainous area is of about $0.3x10^9$ m^3/a and it was implemented deeper. Here the underlying idea is that this component corresponds to the part of the precipitation which infiltrates in the mountainous area through clefts into the deeper rock formations. The groundwater flow system in the bedrock consists of macroscopic structures and cavities linked with each other, as for example fracture networks, faults, layer joints, dissolution features and conduits etc. The groundwater inflow depths were assumed according to the distribution of the water bearing carbonate rocks, sandstones and crystalline rocks in the hydrogeological map of Beijing.

These so-called deep inflows are not dependent directly on the precipitation and change only in terms decades.

The total horizontal groundwater recharge (inflow) ranges from 0.55 to $0.75x10^9$ m^3/a. The quantitative split is to some extend arbitrary and based only on plausibility considerations since none of the components can be measured directly.

2.2.2. Derivation Of Timedependent Groundwater Recharge And Exploitation Data

In order to determine the time-dependent groundwater recharge and exploitation we follow the subsequent procedure:

1) It starts with stating a long term mean value budget for the considered area for getting an idea which fluxes are in which order.

2) In a second step the budget data are regionalized by means of land use maps.

3) Finally the temporal distribution is taken into account by implementing the crop water need (agricultural water demand) and precipitation as dynamic input data such that in the end due to balancing all data, e.g. irrigation, evapotranspiration, etc, become time dependent quantities.

The first task in setting up models covering the water resources of an area of this size is to construct a water budget. The water budget is a theoretical device that supports structuring the water resource system and identifying the most important water fluxes. Here fluxes into and out of the system has to be collected as well as the water fluxes within the model area. The intension must be to realize the relation of water fluxes to each other, to quantify them, separating the more important from negligible water fluxes and to estimate the error that happens due to neglecting them. Since most of the quantities in the water budget are not independent from each other, the quantification of the water budget must be an iterative process. Fig. 5 illustrates the water budget of the region of Beijing. All the before mentioned fluxes are entried. The width of the arrows corresponds to the quantity of the fluxes.

Since the groundwater model is spatially distributed the input data for the groundwater recharge and the exploitation have to be spatially distributed as well. The above mentioned water budget can be regarded as a lumped parameter model for the whole region. The next step is to relate these data to locations by adapting the water balance with respect to specific regions.

- An approach which is followed quite often is to regionalize by means of land use maps. The land use of the considered region is depicted in Fig. 6 showing eleven land use classes. These classes can be summed up to the following four classes:urban and paved areas,
- agriculturally used and irrigated areas,
- water areas and
- non-cultivated areas

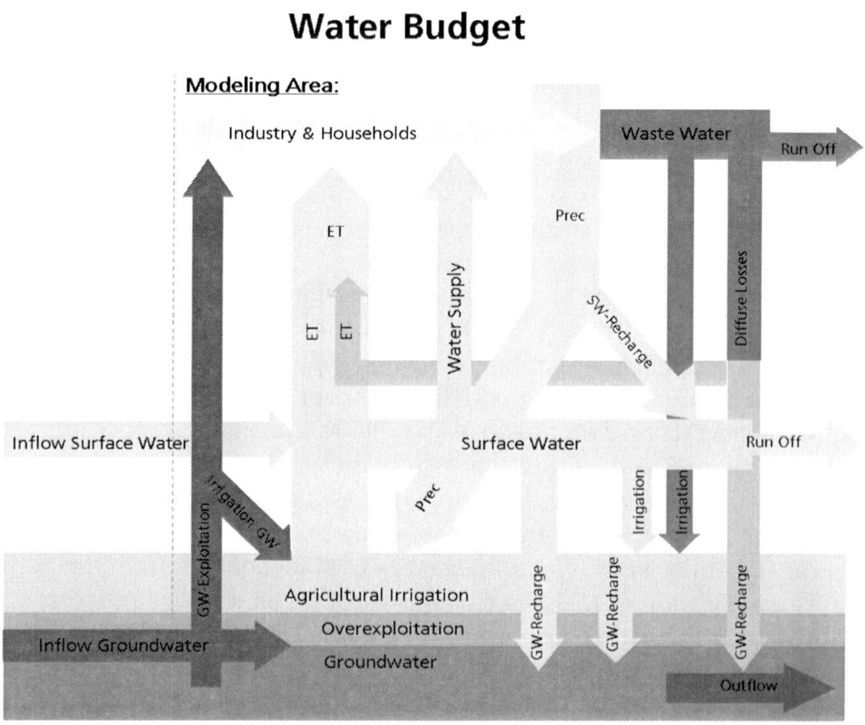

Figure 5. Water budget of the region of Beijing

Regionalization of the water budget means to adapt and evaluate the balance equation

$$P = GWR + ET + SWR \tag{5}$$

for each of these classes. It denotes P the precipitation rate, GWR is the groundwater recharge, ET the evapotranspiration SWR corresponds to the surface water runoff. There are some quantities which are relevant for every class, like infiltration and there are some which are specific for a certain group. And for other classes additional quantities have to be added, i.e. irrigation rates (IRR) for agriculturally used areas.

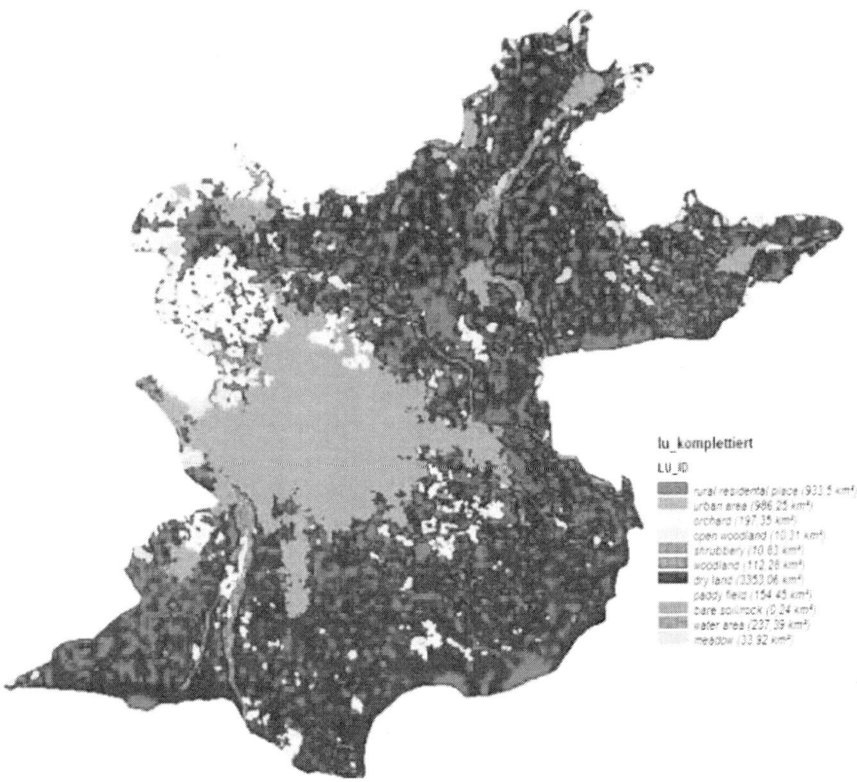

Figure 6. Landuse map of the region of Beijing

2.2.3. Software Realisation Of The Parameter Calculation Within The Dss

The DSS works on the base of so-called *scenarios* whereby a scenario can be understood as set of input data that represents either a real historical situation or an imaginable situation in future with respect to precipitation, exploitation, water use etc. This complete set of input data is applied to the hydrological model to obtain an answer to the question 'What are the effects to the water allocation system'. The historical scenarios usually are used for the validation of the models whereas future scenarios should deliver guidelines for the best way to act in future under certain circumstances. In order to obtain reasonable results from the simulation the scenario has to be complete and consistent. Otherwise the models will gain unreasonable results. A scenario is complete if all data are available the hydrological model expects. In order to keep the system stable often default values are inserted if no data are assigned. But in this case the question is if the data are consistent.

The consistency of data can only be assured by the user. So if we use precipitation rates for groundwater model and for the surface water model for instance in general the information is provided in different data set and formats. In this case the user is responsible, that the data for the groundwater model match the data for the surface water model, i.e. that at same time periods in same regions the same precipitation rates are considered.

Nevertheless, the user can be supported by a parameterization software, as it was realized within the Scenario Wizard of the Beijing DSS to ensure consistency as good as possible. With respect to the groundwater model the scenario is complete and consistent only if the before mentioned data are available for the complete simulation horizon:

- Initial conditions
- Boundary conditions (well fields and inflow)
- Spatially distributed groundwater recharge
- Spatially distributed exploitation

In order to obtain a complete and consistent set of input data for the groundwater model the user has to pass through the groundwater panel of the DSS Scenario Wizard and a so-called groundwater project is created. A groundwater project is a part of the scenario containing all groundwater relevant data.

Within the GW panel the user has to execute five subpanels such that in the end at least a complete set of input data for the groundwater model is generated.

1) The first subpanel calculates the surface water recharge. For this a weighting map is required that determines how much precipitation becomes surface water in direct or indirect manner. On the other hand a reliable precipitation map for the entire model area is necessary. Since the precipitation is not constant over the year we also need temporal weights that define the temporal distribution of the precipitation rate. The Scenario wizard provides a number of precipitation maps of the past which can be also used for future scenarios. The surface water recharge is not a direct input data for the groundwater model but it is required to determine the distribution of exploitation and groundwater recharge in time and space.

2) In the second step the agricultural irrigation from groundwater is and the total spatially distributed exploitation from groundwater is derived. For these calculations a map of the agricultural used areas (irrigated areas) is needed as well as corresponding information about the crop water need/ agricultural water demand. In addition some information about the irrigation from surface water and waste water are requested. Since the agricultural water demand is not constant during the year a temporal distribution is needed as well.

3) In the third subpanel the minimal surface water inflow and the diffuse losses in urban areas is computed and therefore maps of water areas and urban areas are requested. The resulting information is incorporated into the calculation of the groundwater recharge from surface water areas and from urban areas.

4) In this subpanel all input data with respect to the well fields and the inflow parameters have to be defined and entered in to table which asks for yearly data of the exploitation rates from the different well fields and the upper and deeper inflow into the model area.

5) In the last subpanel the generation of the spatial and temporal distribution of the groundwater recharge data is performed. Here the

user again has to provide a weighting map that determines how much of the supplied water becomes groundwater recharge and how much becomes evapotranspiration. The rest of the required data was already determined in the before described subpanels. This step finishes the creation of the groundwater project and all the generated data are documented in a specific groundwater scenario report which is saved within the corresponding project directory. After the simulation of a scenario with a groundwater simulation run a report about simulation results is saved in the project directory as well.

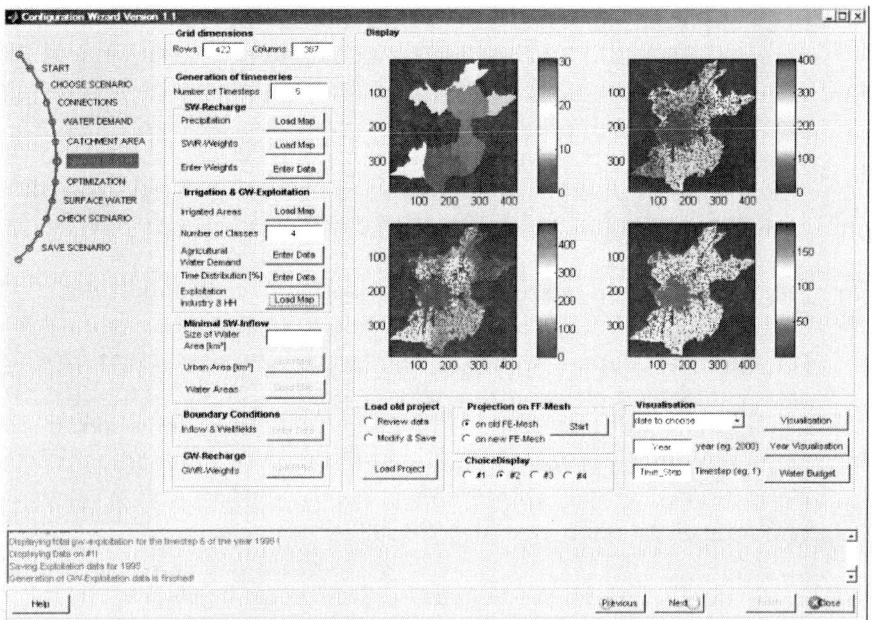

Figure 7. Graphical user interface for the generation of input data like groundwater recharge maps

All computed data are saved as ASCII grid maps which can be read by any GIS system. Figure 7shows the graphical user interface generating groundwater maps.

In order to gain the project data accessible to the FEFLOW simulation model the spatially distributed data have to be projected on the finite element mesh. But in this step no new information is gained therefore it is not a procedure which is important for the scenario generation but it is only a technical requirement.

2.3. Derivation Of Reduced Groundwater Model

For the optimization of the water allocation system the full 3D FEM model (with > 150,000 nodes) is not very suited due to the mentioned large computational time. As for the optimization task a prediction of the hydraulic head (groundwater level) at a set of representative and fixed points is sufficient, an input-output model (e.g. a linear state space model) with considerably smaller order n (e.g. n < 50) than the original FEM model has to be derived.

Methods for model reduction of those large scale systems have gained increasing importance in the last few years [13]. Two main classes of methods for model reduction can be identified, namely methods based on singular value decomposition (SVD) and Krylov based methods. SVD based methods are suited for linear systems and nonlinear systems of an order n < 500 (e. g. balanced truncation for linear systems, proper orthogonal decomposition (POD) for nonlinear systems). Most of these methods have favourable properties like global error bounds and preservation of stability [13]. Krylov based methods are numerically very efficient as only matrix multiplications and no matrix factorization or inversion are needed. Hence they are also suited for large-scale systems. Unfortunately, global error bounds and preservation of stability cannot be guaranteed. Hence actual research is focused on the development of concepts which combine elements of SVD and Krylov based methods [13].

All of these approaches have in common that they aim to approximate the state vector x with respect to a performance criterion, e. g. minimize the deviation between original system and reduced system for a given test input. As a black box input-output model would be sufficient for our purposes there is no need to approximate the whole state vector x. Furthermore, the dimension n of a reduced model which approximates the whole state space vector x would be in most cases n > 100. With this dimension, for the given optimization problem the solution time would be unacceptably high (~ hours). Last but not least the use of the commercial software FEFLOW also prevents the application of e. g. a Krylov based method as no model representation (e. g. state space model) is provided by the software.

The basic idea of the proposed model reduction method is sketched in Fig. 8. We assume the existence of a reference scenario which means that the time dependent input parameters uref(t) of the FEM groundwater model (especially groundwater exploitation Qexpl and recharge Qrech) are determined for the whole optimization horizon. In practical cases these reference scenarios are mostly available or can be generated by plausible assumptions. Hence the task consists in the derivation of a model which approximates the behavior of the full FEM model in the case that the input parameters u differ from uref(t). This model is gained by identification techniques: Test signals (e.g. steps) are added to the reference input uref(t) (dimension p) and the corresponding deviations from the reference output yref(t) (dimension q) are identified. Doing this separately for every component of the input-/output vectors u and y, we finally merge the (p x q) single input-single output (SISO) models to a multi input-multi output (MIMO) model. For the groundwater model, the input parameters are e. g. cumulated (e.g. spatially integrated) exploitation of certain regions or cumulated exploitations of large well fields. The output parameters of the groundwater model are the hydraulic head at representative points ("observation wells"). In our application 13 input and 13 output parameters have been defined by the users: The inputs consist by 9 counties and 4 wellfields, the 13 output parameters are 12 observation wells and the mean hydraulic head of the whole area of the water supply system. As the slow stream groundwater flow can be interpreted as diffusion process (cf. equation (4)) only nearby located input and output parameters (e.g. regions/wellfields and the corresponding observation wells) have some correlation and a SISO model with these input-/output combinations can be gained. Due to this physical reason the number of relevant SISO models is relatively small and hence the resulting MIMO model of relatively low dimension ($n < 50$) which is appropriate for the optimization problem. This proposed approach can be called trajectory and identification based model reduction. A similar approach (for a nonlinear large scale system) is found in [14].

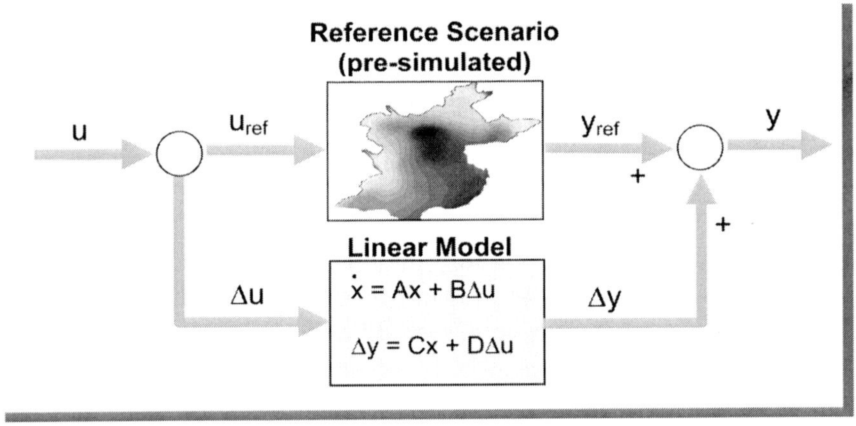

Figure 8. Reduced groundwater model as a linear state space model in combination with a pre-simulated reference scenario.

3. WATER RESOURCES MANAGEMENT AS OPTIMAL CONTROL PROBLEM

The water resources allocation problem is formulated as a discrete-time optimal control problem:

$$\min_{\mathbf{u}^k,\ k=1,\dots,K} \left\{ F\left(\mathbf{x}^K\right) + \sum_{k=0}^{K-1} \mathbf{f}_0^k\left(\mathbf{x}^k, \mathbf{u}^k, \mathbf{z}^k\right) \right\} \tag{6}$$

subject to

$$\mathbf{x}^0 = \mathbf{x}\left(t_0\right) \tag{7}$$

$$\mathbf{x}^{k+1} = \mathbf{f}^k\left(\mathbf{x}^k, \mathbf{u}^k, \mathbf{z}^k\right) \tag{8}$$

$$\mathbf{h}^k\left(\mathbf{x}^k, \mathbf{u}^k, \mathbf{z}^k\right) = \mathbf{0} \tag{9}$$

$$\mathbf{g}^k\left(\mathbf{x}^k, \mathbf{u}^k, \mathbf{z}^k\right) \leq \mathbf{0} \tag{10}$$

The state variables x are the volume content of the reservoirs and channels with small slope and the states of the reduced groundwater model. Control variables u are the discharge of the transport elements as well as the water demand of the customers. The uncontrollable inputs z are the direct precipitation and the potential evaporation for the reservoirs and the flow of rivers entering the considered region, which is derived by means of rainfall-runoff-models. The number of time steps within the optimization horizon is denoted by K.

The process equations (8) consist of the balance equations of the storage nodes and the reduced groundwater model. The balance equations of the non-storage nodes are formulated as general equality constraints (9). The objective function (6) contains the goals of the water management, which are primarily the fulfillment of the customer demand, the compliance with targets for the reservoir and groundwater storage volume and the delivery of water with respect to environmental purposes. Therefore quadratic terms are formulated, which penalize the deviations from desired values, like e.g. for the demand deficit of the demand node j:

$$f_0^k\left(u_j^k\right) = \rho_j^k \Delta t^k \frac{\left(u_j^k - q_{dem,ref,j}^k\right)^2}{{q_{dem,ref,j}^k}^2} \tag{11}$$

where $q_{dem,ref,j}$ is the demand of and u_j is the discharge delivered to the customer. While this term applies for every time step within the optimization horizon, other terms are formulated only for the final point of the horizon, like e.g. for the desired volume content of the reservoirs.

The inequality constraints (10) follow from the technical capabilities of the water distribution system and rules to guarantee safe operation, which are simple bounds for the control variables:

$$\mathbf{u}_{\min}^k \leq \mathbf{u}^k \leq \mathbf{u}_{\max}^k \tag{12}$$

as well as constraints for the reservoir volume xv:

$$\mathbf{x}_{v,\min}^{k} \leq \mathbf{x}_{v}^{k} \leq \mathbf{x}_{v,\max}^{k} \tag{13}$$

and the hydraulic head *hhydr* of the observation wells:

$$\mathbf{h}_{hydr,\min}^{k} \leq g^{k}\left(\mathbf{x}_{gw}^{k}\right) \leq \mathbf{h}_{hydr,\max}^{k} \tag{14}$$

With respect to the practical applicability selected parts of the inequality constraints (10) can be relaxed in order to avoid infeasible optimization problem with respect to unrealistic management demands.

4. NUMERICAL SOLUTION OF THE OPTIMAL CONTROL PROBLEM

The optimal control problem is numerically solved as large scale structured non-linear programming problem (NLP):

$$\min_{\mathbf{y}} \left\{ J(\mathbf{y}) \mid \mathbf{h}(\mathbf{y}) = \mathbf{0}; \mathbf{g}(\mathbf{y}) \leq \mathbf{0} \right\} \tag{15}$$

The optimization variables are the state and control variables of the several stages in time:

$$\mathbf{y} = \left[\left(\mathbf{x}^{0}\right)^{\mathrm{T}} \left(\mathbf{u}^{0}\right)^{\mathrm{T}} \ldots \left(\mathbf{x}^{K-1}\right)^{\mathrm{T}} \left(\mathbf{u}^{K-1}\right)^{\mathrm{T}} \left(\mathbf{x}^{K}\right)^{T} \right]^{T} \tag{16}$$

The state equations of the process model are incorporated as equality constraints:

$$\mathbf{h}(\mathbf{y}) = \begin{bmatrix} \mathbf{x}^{0} - \mathbf{x}(t_{0}) \\ \mathbf{f}^{0}\left(\mathbf{x}^{0}, \mathbf{u}^{0}, \mathbf{z}^{0}\right) - \mathbf{x}^{1} \\ \vdots \\ \mathbf{f}^{K-1}\left(\mathbf{x}^{K-1}, \mathbf{u}^{K-1}, \mathbf{z}^{K-1}\right) - \mathbf{x}^{K} \end{bmatrix} \tag{17}$$

The advantage of this problem formulation with *(K(n+m)+n)* optimization variables is the special sparsity structure with a block-diagonal Hessian-matrix and block-banded Jacobian matrices (*k*: number of time steps, *n*: number of state variables, *m*: number of control variables), which follows from the fact, that the process equations as coupling element between adjusting stages are linear in the state variables x^{k+1}. The numerical solution of this problem requires about $(K(n+m)^3)$ basic arithmetic operations.

One alternative approach consists of eliminating the state variables from the optimization problem. The reduced dimension of the according non-linear programming problem with $(K\ m)$ optimization variables comes along with a loss of structure. The Hessian and Jacobian matrices are full. Because of the solution effort of order $(K\ m)^3$ and the large amount of control variables (number of edges in the network description) this approach is not promising for this special field of application.

For the numerical solution of large scale non-linear programming problems interior point (IP) solver have become popular during the last years because of their superior behavior for NLPs with many inequality constraints. In this approach the objective function is expanded by adding barrier terms for the inequality constraints:

$$\min_{\mathbf{y}} \left\{ J(\mathbf{y}) + \mu \sum_{j=1}^{n_g} \ln\left(-\mathbf{g}_j(\mathbf{y})\right) \mid \mathbf{h}(\mathbf{y}) = \mathbf{0} \right\} \tag{18}$$

The solution of the original NLP (15) results from the subsequent solution of (17) with a decaying sequence of $\mu \to 0$. The identification of right active set with its combinatorial complexity is avoided. The state of the art non-linear interior point solver IPOPT is used for the application at hand [15]. The interface for multistage optimal control problems of the optimization solver HQP [16], which provides an efficient way for problem formulation along with routines for the derivation of ∇J, ∇h, ∇g by means of automatic differentiation [17], is used and coupled to IPOPT.

A typical water management problem (horizon of five years, discretization of one month) has about 8000 optimization variables, 5500 equality constraints and 7200 inequality constraints. The numerical solution takes approximately 60 iterations and a calculation time of 10 seconds on an Intel Core 2 Duo CPU (2.5 GHz). Table 1 shows the linear dependency of the numerical solution effort from the number of time steps within the optimization horizon.

Table 1. Numerical solution effort in dependence on the optimization horizon

optimization horizon	Numerical solution effort			
	optimization variables	Iterations	main storage [MB]	calculation time [s]
30days	7898	56	52	9.6
10 days	23880	66	145	35.1
5 days	47853	56	286	61.7

5. FIRST RESULTS OF THE OPTIMAL WATER MANAGEMENT APPROACH

The proposed concept for optimal water management is applied to the Beijing region. The region has precipitation, which varies geographically, seasonally and yearly. Eighty-five percent of rainfall falls between July and September. Groundwater is the most important source of water for the Beijing region, covers about 50-70%. Beijing has suffered from over exploitation of this source over the years. Surface water supply in the Beijing region depends mainly on upstream inflows of the major river systems Chaobai, North Grand Canal and Yongding. Aside from problems such as excessive withdrawal and water quality deterioration of surface waters, the lack of regional coordination leads to issues such as uncoordinated withdrawals. Besides these problems, the water supply system is subject to other common problems, such as rapid population growth and urbanization, decentralized reservoir/groundwater management, changing attitude towards sustainability and attribution to greater attention of environmental issues.

The optimal water management system is evaluated for three scenarios based on the same set of input data, which are a combination of historical rainfall measurements and customer demands reflecting the predicted development of population size and economic growth. The objective function contains four quadratic terms in order to penalize deviations from the desired final water level of largest reservoir in the system (Miyun reservoir), from the desired final average hydraulic head of the groundwater storage as well as the deficit of the customer demand separated into two groups for household/industry and agriculture. The deficit of the delivered water must be less than 5 % for domestic/industrial clients and less than 25 % for agricultural clients. For the third scenario it is assumed that water can be transferred to the considered region up to an annual amount of 300 Mio. m^3 starting from the third year within the optimization horizon.

The overall water demand exceeds noticeable the natural sources. The reduced groundwater model has been derived under the assumption that this over-consumption was covered by the groundwater storage. This leads to a strong reduction of the average hydraulic head of the groundwater storage of about 9 meters during 5 years (scenario 1, see Fig. 10). Using the initial state of the Miyun reservoir and the final value of the reference trajectory for the average groundwater head as target in the objective function, for the base scenario (scenario 1) there are only small deviations from these values in combination with a minor demand deficit observable. In the second scenario the increase of the target value for the final groundwater hydraulic head at 5 m and a corresponding shift of penalty coefficients in order to keep this value lead to a better spreading of the overdraft over the different storages as well as to the customers.

Fig. 9 shows the course of the water level for two reservoirs. Because of its capacity of 4.4×10^9 m^3 the Miyun reservoir plays an important role for the long term management of the overall system. As can be seen, the years with above-average precipitation produce only a medium rise of the water level. The second scenario, which attempts to reduce the decline of the average hydraulic head of the groundwater storage, results in an increased release of this reservoir compared to the base scenario, which corresponds to a change of the final water level from 142 m above sea level to 137 m. In the third scenario a part of this release is replaced by water from outside of the considered region, which reduces the water level decrease at about 2 m. The second reservoir is situated at a channel from the Miyun-reservoir to the city of Beijing and serves only as intermediate storage. The admissible range for water management is completely utilized by the optimal control approach. The shift in the management target causes a different operating strategy because water from the connected channel is taken to replace groundwater abstractions in the nearby regions. The change of the management target for scenario 2 and scenario 3 induce also an increase of the overall demand deficit from 0.1 % to up to 3.5 % (scenario 2), which is nearly 6.8×10^8 m^3 over the full horizon.

Fig. 10 shows the time plot of the mean groundwater level of the optimization scenarios. It is obvious that in scenarios 2 and 3 the aimed increase of the target value for the final groundwater hydraulic head at 5 m is nearly achieved. In Fig. 11 the impact of the different strategies to the exemplary input 1 (exploitation in a certain region) and exemplary output 5 (groundwater level at a defined observation point) can be studied. The exploitation is clearly decreased in scenarios 2 and 3 which correspond to an increase of the groundwater level at the observation point. Finally, from Fig. 12 it can be seen that the performance of the drastically reduced groundwater model is good, reflecting the fact that the original FEM model with more than 100,000 nodes has been reduced to a state space model with 36 states.

6. CONCLUSIONS

In this paper an optimal control approach as a component of a decision support system (DSS) for the management of the total water resources (surface water, groundwater, external water resources) in a fast developing region under a critical water shortage has been presented. It has been proven that in spite of the large area which has to be managed and the corresponding complex surface and groundwater models the optimization problem could be solved in an appropriate computation time (~ minutes). This could be achieved by a drastic reduction of the complex groundwater model to a state space model of relatively low dimension ($n < 50$). The user (i.e. water allocation decision maker) is enabled to select from a number of predefined performance criteria as well as to assign constraints to the elements of the water allocation system in order to specify the management targets according to his/her needs. The performance of the proposed concept is demonstrated by close to reality optimization scenarios, whereby the benefit of a new strategic channel has been investigated with a planning horizon of 5 years. Actually the developed DSS component is used in a first version by the decision makers. Future work will focus on the application and adaptation of the developed concept and software for the water resources management of further regions with critical water shortage.

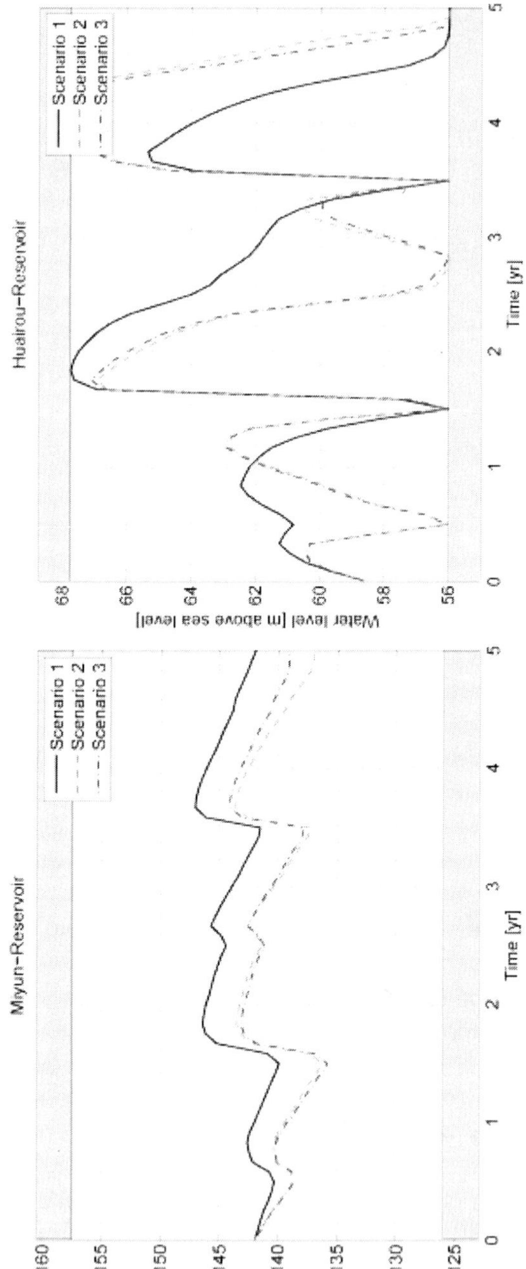

Figure 9. Water level of the largest reservoir in the water distribution system (Miyun-reservoir) as well as of a reservoir with seasonal storage capacity (Huairou-reservoir).

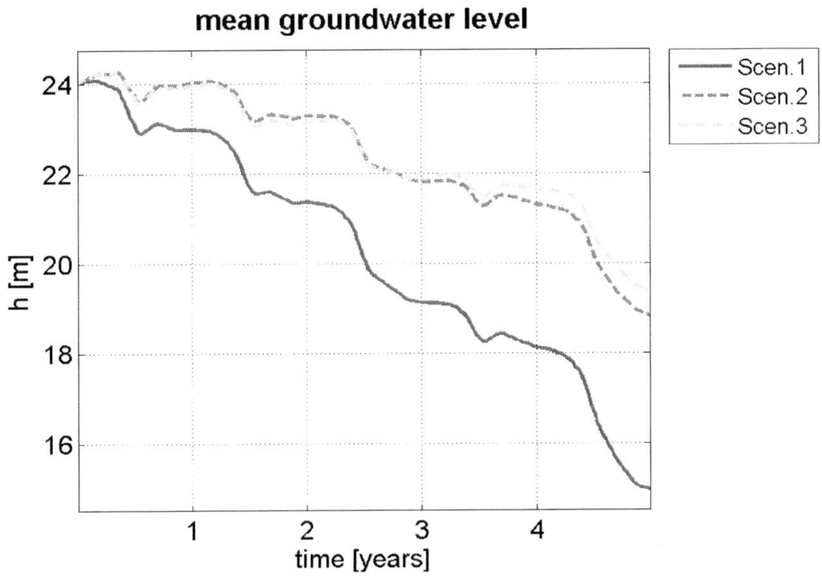

Figure 10. Time plot of the mean groundwater level in the optimization scenarios 1 – 3.

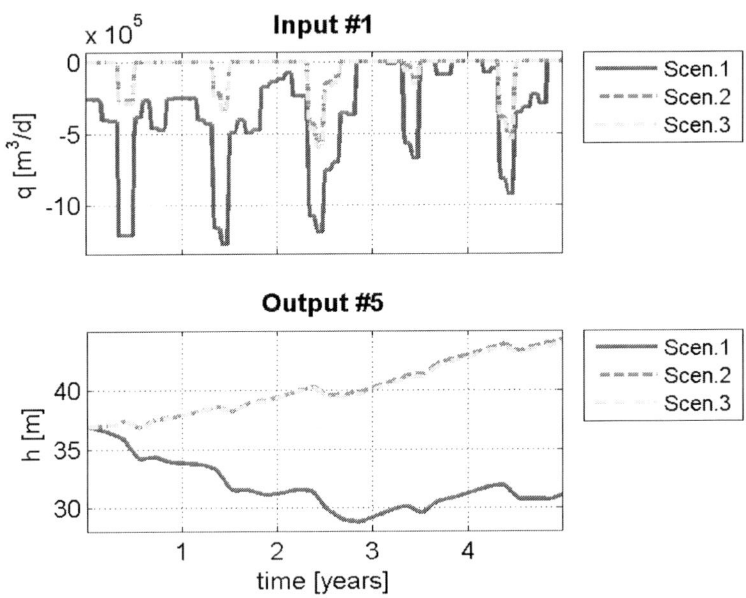

Figure 11. Time plot of input 1 (exploitation in a certain region) and output 5 (groundwater level at a defined point) in the optimization scenarios 1 – 3.

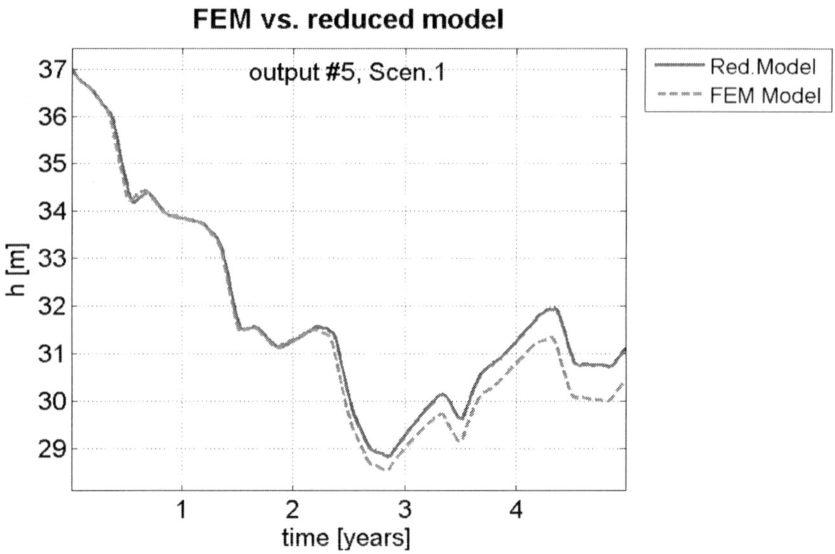

Figure 12. Time plot of output 5 of the full FEM model compared with the reduced model (optimization scenario 1).

ACKNOWLEDGEMENTS

This work was supported by the German Federal Ministry of Education and Research (BMBF).

REFERENCES

1. Bernard T., Krol O., Linke H., Rauschenbach T. Optimal Management of Regional Water Supply Systems Using a Reduced Finite-Element. at-Automatisierungstechnik, 2009; (57)12: 593-600.

2. Rauschenbach T., Steusloff H., Birkle M. Modeling of Beijing municipality's water resources as basis for an optimal water allocation system. In: Proceedings of Beijing Conference on Sustainable Water Management. Beijing, China; 2007.

3. Mays L. W. Water resources system management tools. New York: McGraw-Hill Companies; 2005.

4. McKinney D. C. International Survey of Decision Support Systems for Integrated Water Management. Technical Report, IRG Project No. 1673-000, Bucharest, 2004.

5. Rauschenbach T.WaterLib 2.0 - Simulation and Management of Surface WaterSystems. In: 2008 Promotion Conference on International Advanced Water Technology.Beijing, China, 2008.

6. Bear J. Dynamic of Fluids in Porous Media, New York, Dover Publications; 1988.

7. Software package FEFLOW: http://www.feflow.info/ (accessed 5 June 2012)

8. Bernard T., Linke H., Krol O. A Concept for the long term Optimization of regional Water Supply Systems using a reduced Finite Element Groundwater Model, VDI-Berichte 1980, Düsseldorf: VDI-Verlag, 2007; p751-762

9. ChenJ.Y., TangC.Y., ShenY.J., SakuraY., KondonA.,ShimadaJ. Use of water balance calculationand tritium to examine the dropdown of groundwatertable in the piedmont of the north china plain (ncp). Environmental Geology, 44:564-571, 2003.

10. Z. Han. Groundwater resources protection andaquifer recovery in china. Environmental Geology,44:106-111, 2003.

11. Zippel M., Hannappel S. Ermittlung des Grundwasserdargebotes der Berliner Wasserwerke mittels regionaler numerischer Grundwasserströmungsmodelle. Grundwasser, 4:195-207, 2008.

12. FosterS., GardunoH., EvansR., OlsonD., TianY.,ZhangW., HanZ. Quarternary aquifer of thenorth china plain. Hydrogeology Journal, 12:81-93,2004.

13. Antoulas A. C. Approximation of Large-Scale Dynamical Systems. Philadelphia: SIAM Press, 2005.

14. Wolfrum P., Vargas A., Gallivan M., Allgöwer F.: Complexity reduction of a thin film deposition model using a trajectory based nonlinear model reduction technique. American Control Conference, Portland, USA, 2005.

15. Wächter A., Biegler L. T. On the implementation of a primal-dual interior point filter line search algorithm for large-scale nonlinear programming. Mathematical Programming, 2006: 106(1):25–57.

16. Franke R., Arnold E. The solver Omuses/HQP for structured large-scale constrained optimization: algorithm, implementation and example application, Sixth SIAM Conference on Optimization, Atlanta, USA, 1999.

17. Griewank A., Juedes D., Utke J. ADOL-C: A package for the automatic differentiation of algorithms written in C/C++. ACM Trans. Math. Software, 1996: 22(2): 131-167.

Chapter 4

A Review of Stakeholder Participation in Watershed and Urban Water Supply Management

Nur Asikin Roslan[1], **Noorul Hassan Zardari**[2], **Irena Naubi**[1]

[1]Department of Hydraulics and Hydrology, Faculty of Civil Engineering, Universiti Teknologi Malaysia, Skudai, Johor Malaysia

[2]Institute of Environmental and Water Resources Management (IPASA), Universiti Teknologi Malaysia, Skudai, Johor Malaysia

ABSTRACT

Stakeholder participation had been recognized as one of the useful methods to be embedded in decision making process. Stakeholder can be defined as group of people who can affect and be affected by the decision. Thus there is a need of approach which can address their best interests in any problem which can affect or bring benefit to them. However applications of this approach are still fragmented in watershed and urban water supply management. Thus, this paper tries to explore the literature to see the growing importance of stakeholder participation concept that had been applied in watershed and urban water supply management projects for a selected period of thirteen years (2000-2012). For that, we searched four main academic databases namely sciencedirect, scopus, springer and web of knowledge for three different sets of keywords 'stakeholder participation', 'watershed management', and 'urban water supply management'. The search was made to see the 'application' and 'non application' publications in journals only for thirteen years span. The search findings were sorted out based on the year of publications, country of the origins of the author institutions and journal title in which they were published. This study may also serves as reference for decision maker on stakeholder participation approach in urban water supply management.

KEYWORDS

Stakeholder Participation, Watershed Management, Urban Water Supply Management, Review

1. INTRODUCTION

Research conducted by[1] on stakeholder defines stakeholder as parties that can be affected by the decision or situation created by the decision. However, in the modern context, stakeholder can also affect the decision that yet to be taken as the recognizing role of stakeholder in the process. This importance of stakeholder had been elaborated in quite number of works such as[2],[3] and[4].

In urban water supply management, stakeholder participation had been recognized as a useful tool to be integrated in the policy making and management process. This follows after recommendation Agenda 21 in United Nations Conference on Environment and Development (UNCED) that the broadest public participation and the active involvement of the non-governmental organizations and other groups should be encouraged. The process had been recommended to be included as early as possible in decision making cycle[5]. It also calls for integration of environment and economics in decision making which is new at that time since environmental consequences are often left behind in pursuing economic benefit which had been addresses in few researches such as by[6] and[7] Despite its vast use in other sectors such as fisheries[8], [9],[10]; agriculture[3],[11]; business practice, healthcare, energy, environment and private sector project, the application in urban water supply management is still few.

This paper tries to explore the growing importance of stakeholder participation in urban water and urban water supply management, the degree of participation and the method of participation. Even such review has been done in other such as in environmental management[12], European based water policy[13] and ecological risk assessment[14] and the review of stakeholder participation in general watershed and water supply management field are still missing. This review will serve as platform to gather information and hope to serve as guideline for decision makers in urban water supply for inclusion of the stakeholder participation in the decision process.

2. DATABASE SEARCH RESULTS

This study relies heavily on secondary data from databases. For purpose of reviewing past researches, databases including ScienceDirect, Springer, Scopus and Web of Knowledge were searched for publications regarding the stakeholder involvement in watershed management. Criteria used for databases survey is strictly following certain aspect such as by using same keyword search that is 'stakeholder' and 'watershed and water supply' throughout databases. Result then is taken from specific years that is from year 2000 until 2012. Only journal publications were considered for purpose of this survey excluding book chapter. The results were then sorted out according to year of publication.

It can be observed in Figure 1 that ScienceDirect gives the most number of publications followed by Springer, Scopus. Web of Knowledge database gives least number of publications for the selected period. Trend of publishing increases by years which see sharp increase of total publications between the year 2006 to 2007 with 81 publications and also between year 2011 and 2012 increment of 102 publications. This may be due to introduction of socio-economic aspect in Integrated Water Resource Management (IWRM). In 1996, Global Water Partnership (GWP) is established to foster IWRM which aims to ensure coordinated development and management of water, land, and related resources by maximizing economic and social welfare in an equitable manner without compromising the sustainability of vital ecosystems[15]. In GWP Framework for Action which is established in the 2000 Water Forum in The Hague, it called for encouragement of good governance so that involvement of public and interest of stakeholders are included in the management of water resources. This also mark the transition of management based on the

Water Demand Management (WDM) to IWRM. Since then, number of studies on stakeholder participation in Water Resource Management is increasing.

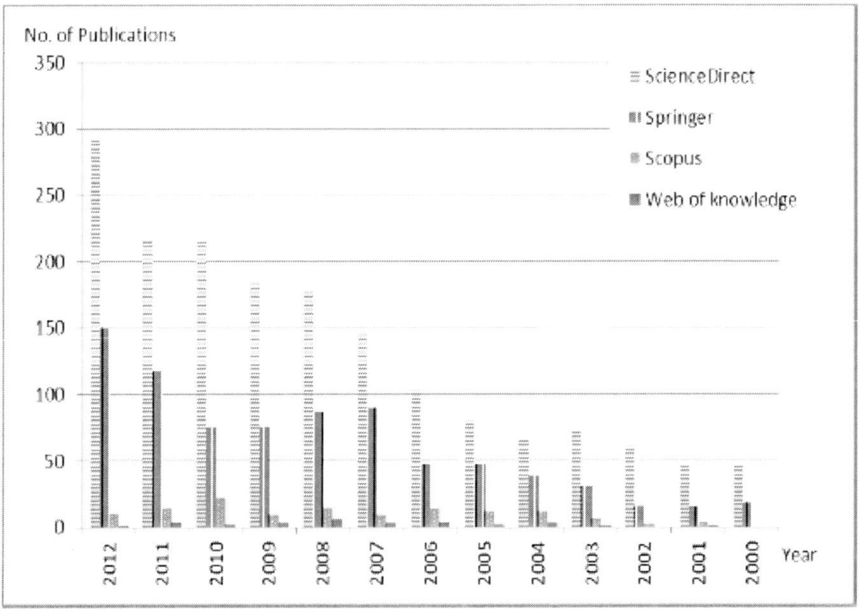

Figure 1. Number of papers published in different databases in various years

3. ANALYZING RELEVANT PUBLICATIONS

Early survey only shows rough estimation to the real number of relevance publications. The databases were further sorted out based on the relevance to the research area of this review. The sorting were including year 2013 to look into new trend of participation documentation of these publications. The total publications from ScienceDirect and Springer obtained by the database search were go through one by one to roughly see the method used, the degree of participation and country of origin. A total of 151 journals papers were found to be relevant with this study. The total 151 publications were compared across the year, country of origin of author institutions and also application and non-application in watershed and water supply management area.

When sorting out the journal publications from databases, other publication deemed relevant to the research were also taken into account where similar keywords such as 'social learning', 'public engagement' and 'stakeholder consultation' method are used in the publications. This, even if the method type are slightly different than participation and where objectives of such method are different, were included. Further exploration and explanation on the differences of each method is not included in the scope of this review. Figure 2 below show relevant publications distribution according to year.

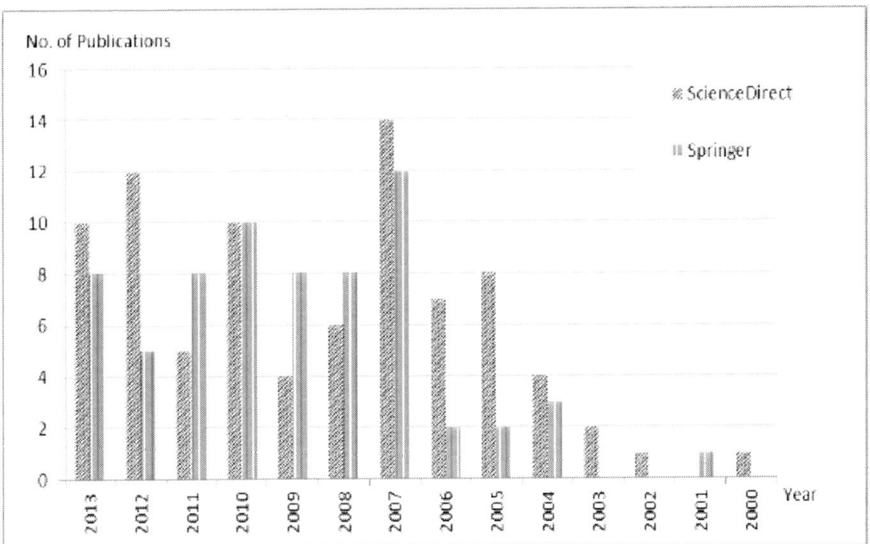

Figure 2. Papers published in ScienceDirect and Springer databases on yearly basis

Overall, the total relevant papers from ScienceDirect are more that is 84 papers than Springer which has 67 papers. Relevant papers also shows different distribution for research according to year where the year with highest total publication is in 2007 with 27 publications, different from preliminary keywords survey from databases. Second year with high publication number is in 2010 that is 21 papers.

Country wise distribution shows that the United States are on the top of publishing countries with 27 papers followed by the United Kingdom with 19 paper. Australia with 14 publications and Germany with 13 publications rank 3rd and 4th respectively. While top scorer countries are mostly consist of developed countries, developing countries fill certain spot in least publications countries such as South Korea, India, Ghana, Ethiopia, Lebanon, Mozambique, Laos and Syria. The distributions are shown as in Figure 3 below. The papers were further sorted out based on the journal title and separated between that from Science Direct and Springer as shown in Table 1 which gives early prediction on the field that this method is likely used for.

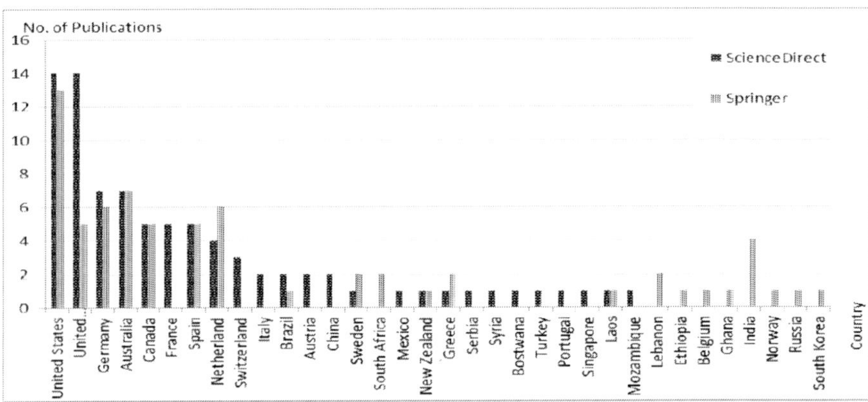

Figure 3. Comparison of publication according to country of origin of the paper first author institution between ScienceDirect and Springer database

In this paper, we also distributed the database search results into 'applications' and 'non application' of publications. In case of 'applications', paper which presented case studies dealing with solving any real-world problems associated to watershed, urban water supply management along with the search keyword of 'stakeholder participation' are grouped together. On the other hand, papers which mainly focused on methodology and mathematical derivations on the stakeholder participation in watershed and urban water supply management were put together. Both this 'application' and 'non-application' paper were listed in Appendix A.

Table 1. Sorting of publications based on journal title and database

ScienceDirect		Springer	
Journal Title	No.	Journal Title	No.
Environmental Modelling And Software	12	Water Resource Management	28
Journal Of Environmental Management	8	Environmental Management	14
Environmental Science And Policy	8	Regional Environmental Change	6
Ecological Economics	7	Climate Change	4
Physics And Chemistry Of The Earth	6	Environment, Development and Sustainability	3
Journal Of Hydrology	6	Urban Forum	2
Land Use Policy	5	Other Publications (only one paper published in each journal)	10
Utility Policy	3	**TOTAL**	67
World Development	2		
Global Environmental Change	2		
Agricultural Water Management	2		
Desalination	2		
Geoforum	2		
Journal Of Cleaner Production	2		
Other Publications (only one paper published in each journal)	14		
TOTAL	84		

The papers published in 'application' category were further divided into five types of different applications in solving some real-world problems where stakeholders were engaged and their perceptions considered while planning or management for resources. These five different areas of applications include: water resources management, watershed management or river basin management, integrated water resources management (IWRM) or integrated river basin management (IRBM), water governance and water rights, water supply management and others. Table 2 shows the distribution of applied paper according to their field.

Table 2. Paper distribution according to field of application

Application Field	References
Water Resource Management	[19] [20] [33] [36] [39] [43] [50] [51] [52] [54] [55] [70] [74] [76] [87] [91] [92] [99] [100]
Watershed Management or river basin management	[15] [16] [22] [26] [38] [46] [49] [57] [61] [64] [65] [71] [75] [88] [90] [96] [97] [102] [103] [105]
Integrated Water Resource Management (IWRM) or Integrated River Basin Management	[17] [18] [21] [21] [25] [30] [40] [47] [59] [72] [77] [80] [81] [86] [89] [95] [98]
Water governance and water rights	[29] [34] [35] [42] [45] [58] [62] [63] [69] [78]
Water Supply management	[24] [27] [31] [32] [37] [41] [53] [66] [79] [85] [93] [101]
Miscellaneous (irrigation and agriculture, water utility, groundwater; landuse, etc.)	[28] [44] [48] [56] [60] [67] [73] [82] [83] [84] [94] [104]

4. CONCLUSIONS

The growing importance of stakeholder participation had successfully been demonstrated in this paper by means of database survey that shows increasing publications from year 2000 to 2012. This growing number motivates further researches on relevant publications which later includes year of 2013, to get into insights of the ideas and trends of the participation studies that had been published. Thus, total numbers of 151 publications relevant to this review were sorted out for detail analysis. ScienceDirect give the most number of publications followed by Springer, Scopus and Web of Knowledge. The results agree that the publications trend is increasing and this is not only just numbers but also the ideas and theories underlying the process which is demonstrated in the publications of application and non-application paper.

Country wise, the United States come up with most number of publications followed by the United Kingdom, Australia and Germany while according to field of applications, watershed management and water resource management appear with most number of publications compared to other fields such as water supply, and Integrated Water Resource Management (IWRM).

This overall review can give slightest idea on how the field had expand in last 13 years and also may be serves as guideline in stakeholder participation in watershed and urban water supply management.

ACKNOWLEDGEMENTS

This study was funded by the Ministry of Higher Education (MOHE) Malaysia and the Universiti Teknologi Malaysia (UTM) under GUP Grant Tier-2 with Vot No. 07J27.

REFERENCES

1. Burger, J. Stakeholder and scientist: achieving implementable solutions to energy and environmental issues, Springer.,2011.

2. Few, R., Brown K. and Tompkins, E.L., 2007, Public participation and climate change adaptation: avoiding the illusion of inclusion. Climate Policy. 7: 46-59

3. Neef A. and Neubert D., 2011, Stakeholder participation in agricultural research projects: a conceptual framework for reflection and decision-making. Agriculture and Human Value. 28: 179-194

4. Hage M., Leroy P. and Petersen A.C., 2010, Stakeholder participation in environmental knowledge production. Future.42: 254-264

5. Starkl M., Brunner N.,Flogl W., Wimmer J., 2009, Design of an institutional decision-making process: The case of urban water management. J. of Env. Management.90: 1030-1042

6. Nitirach.S-N, and Vilas N., 2011, Strategic decision making for urban water reuse application: A case from Thailand. Desalination.268 (1-3): 141-149

7. Okeola O.G. and Sule B.F., 2011, Evaluaiton of management alternatives for urban water supply sytem using Multi Criteria Decision Analysis.Journal of King Saud University – Engineering Sciences. 24: 19-24.

8. Varjopuro R., Gray T., Hatchard J., Rauschmayer F. and Wittmer H. 2008. Introduction: Interaction between environment and fisheries—The role of stakeholder participation. Marine Policy. 32: 147-157.

9. Gray T. and Hatchard J., 2008, A complicated relationship: Stakeholder participation and the ecosystem-based approach to fisheries management. Marine Policy. 32: 158-168

10. Berghöfer, A., Wittmer, H. and Rauschmayer, F., 2008, Stakeholder participation in ecosystem-based approaches to fisheries management: A synthesis from European research projects. Marine Policy. 32: 243-253.

11. ZardariN.u.H, Cordery I. and Sharma A., 2010, An objective multi attribute analysis approach for allocation of scarce irrigation water resources. Journal of American Water Resources Association. 46(2): 412-428

12. Stinger L.C., Dougil A.J., Fraser E.,Hubback K., Prell C. and Reed M.S., 2006, Unpacking "Participation" in adaptive management of social-ecological systems: a critical review. Ecology and Society. 11(2): 39

13. Videira N., Antunes P., Santos R. and Lobo G., 2006, Public and stakeholder participation in European water policy: a critical review on project evaluation process. European Environment. 16: 19-31

14. Glivken J., 2000, Getting stakeholder participation 'right': a discussion of participatory process and possible pitfalls. Environmental Science and Policy. 2: 305-310

15. GWP, 2000, Integrated Water Resources Management. Global Water Partnership, Technical Advisory Committee (TAC) Background Papers No. 4, Stockholm.

16. Ananda, J. and W. Proctor, 2013, "Collaborative approaches to water management and planning: An institutional perspective." Ecological Economics 86(0): 97-106.

17. Arnette, A., C. Zobel, et al., 2010, "Stakeholder ranking of watershed goals with the vector analytic hierarchy process: Effects of participant grouping scenarios." Environmental Modelling & Software 25(11): 1459-1469.

18. Barthel, R., S. Janisch, et al., 2008, "An integrated modelling framework for simulating regional-scale actor responses to global change in the water domain." Environmental Modelling & Software 23(9): 1095-1121.

19. Benito, G., R. Rohde, et al., 2010, "Management of Alluvial Aquifers in Two Southern African Ephemeral Rivers: Implications for IWRM." Water Resources Management 24(4): 641-667.

20. Berg, H. and T. BenDor, 2010, "A Case Study of Form-Based Solutions for Watershed Protection." Environmental Management 46(3): 436-451.

21. Berke, P., D. Spurlock, et al., 2013, "Local comprehensive plan quality and regional ecosystem protection: The case of the Jordan Lake watershed, North Carolina, U.S.A." Land Use Policy 31(0): 450-459.

22. Bhave, A., A. Mishra, et al., 2013, "Sub-basin scale characterization of climate change vulnerability, impacts and adaptation in an Indian River basin." Regional Environmental Change: 1-12.

23. Brinkman, E., E. Seekamp, et al., 2012, "Community Capacity for Watershed Conservation: A Quantitative Assessment of Indicators and Core Dimensions." Environmental Management 50(4): 736-749.

24. Broad, K., A. Pfaff, et al., 2007, "Climate, stream flow prediction and water management in northeast Brazil: societal trends and forecast value." Climatic Change 84(2): 217-239.

25. Brunner, N., A. Lele, et al., 2010, "Water sector reform policy of India: Experiences from case studies in Maharashtra." Journal of Policy Modeling 32(4): 544-561.

26. Cash, C. and L. Swatuk, 2011, "Integrated Development Planning in South Africa: Lessons from the Dwars River Valley." Urban Forum 22(1): 53-73.

27. Ceccato, L., V. Giannini, et al., 2011, "Participatory assessment of adaptation strategies to flood risk in the Upper Brahmaputra and Danube river basins." Environmental Science & Policy 14(8): 1163-1174.

28. Chanan, A. and P. Woods ,2006, "Introducing total water cycle management in Sydney: a Kogarah Council initiative." Desalination 187(1–3): 11-16.

29. Cinderby, S., A. d. Bruin, et al., 2011, "Participatory geographic information systems for agricultural water management scenario development: A Tanzanian case study." Physics and Chemistry of the Earth, Parts A/B/C 36(14–15): 1093-1102.

30. Clark, J. A. and C. Semmahasak. 2013, "Evaluating Adaptive Governance Approaches to Sustainable Water Management in North-West Thailand." Environmental Management 51(4): 882-896.

31. Cohen, S., D. Neilsen, et al., 2006, "Learning with Local Help: Expanding the Dialogue on Climate Change and Water Management in the Okanagan Region, British Columbia, Canada." Climatic Change 75(3): 331-358.

32. Dawadi, S. and S. Ahmad, 2013, "Evaluating the impact of demand-side management on water resources under changing climatic conditions and increasing population." Journal of Environmental Management 114(0): 261-275.

33. de Graaf, R. and R. v. der Brugge, 2010, "Transforming water infrastructure by linking water management and urban renewal in Rotterdam." Technological Forecasting and Social Change 77(8): 1282-1291.

34. Deng, M., D. Qin, et al., 2012, "Public perceptions of climate and cryosphere change in typical arid inland river areas of China: Facts, impacts and selections of adaptation measures." Quaternary International 282(0): 48-57.

35. Dore, J. and L. Lebel, 2010, "Deliberation and Scale in Mekong Region Water Governance." Environmental Management 46(1): 60-80.

36. Dore, J., L. Lebel, et al., 2012, "A framework for analysing transboundary water governance complexes, illustrated in the Mekong Region." Journal of Hydrology 466–467(0): 23-36.

37. Doummar, J., M. A. Massoud, et al., 2009, "Optimal Water Resources Management: Case of Lower Litani River, Lebanon." Water Resources Management 23(11): 2343-2360.

38. Farolfi, S., J.-P. Müller, et al., 2010, "An iterative construction of multi-agent models to represent water supply and demand dynamics at the catchment level." Environmental Modelling & Software 25(10): 1130-1148.

39. Floress, K., L. Prokopy, et al., 2011, "Who's in Charge: Role Clarity in a Midwestern Watershed Group." Environmental Management 48(4): 825-834.

40. Gaddis, E. J. B., H. H. Falk, et al., 2010, "Effectiveness of a participatory modeling effort to identify and advance community water resource goals in St. Albans, Vermont." Environmental Modelling & Software 25(11): 1428-1438.

41. Gallego-Ayala, J. and D. Juízo, 2012, "Performance evaluation of River Basin Organizations to implement integrated water resources management using composite indexes." Physics and Chemistry of the Earth, Parts A/B/C 50–52(0): 205-216.

42. Garfi, M., L. Ferrer-Martí, et al., 2011, "Multi-criteria analysis for improving strategic environmental assessment of water programmes. A case study in semi-arid region of Brazil." Journal of Environmental Management 92(3): 665-675.

43. Garrick, D., M. A. Siebentritt, et al., 2009, "Water markets and freshwater ecosystem services: Policy reform and implementation in

the Columbia and Murray-Darling Basins." Ecological Economics 69(2): 366-379.

44. Georgakakos, A. P., H. Yao, et al., 2012, "Value of adaptive water resources management in Northern California under climatic variability and change: Reservoir management." Journal of Hydrology 412–413(0): 34-46.

45. Grolleau, G. and L. M. J. McCann, 2012, "Designing watershed programs to pay farmers for water quality services: Case studies of Munich and New York City." Ecological Economics 76(0): 87-94.

46. Hammer, M., B. Balfors, et al., 2011, "Governance of Water Resources in the Phase of Change: A Case Study of the Implementation of the EU Water Framework Directive in Sweden." AMBIO 40(2): 210-220.

47. Harmancioglu, N. B., K. Fedra, et al., 2008, "Analysis for sustainability in management of water scarce basins: the case of the Gediz River Basin in Turkey." Desalination 226(1–3): 175-182.

48. Hartmann, J., J. Levy, et al., 2006, "Managing Surface Water Contamination in Nagoya, Japan: An Integrated Water Basin Management Decision Framework." Water Resources Management 20(3): 411-430.

49. Henriques, C., I. P. Holman, et al., 2008, "An interactive multi-scale integrated assessment of future regional water availability for agricultural irrigation in East Anglia and North West England." Climatic Change 90(1-2): 89-111.

50. Hermans, C., J. Erickson, et al., 2007, "Collaborative environmental planning in river management: An application of multicriteria decision analysis in the White River Watershed in Vermont." Journal of Environmental Management 84(4): 534-546.

51. Hommes, S., J. Vinke-de Kruijf, et al., 2009, "Knowledge and Perceptions in Participatory Policy Processes: Lessons from the Delta-Region in the Netherlands." Water Resources Management 23(8): 1641-1663.

52. Iglesias, A., L. Garrote, et al., 2011, "Re-thinking water policy priorities in the Mediterranean region in view of climate change." Environmental Science & Policy 14(7): 744-757.

53. Isendahl, N., A. Dewulf, et al., 2010, "Making framing of uncertainty in water management practice explicit by using a participant-structured approach." Journal of Environmental Management 91(4): 844-851.

54. Ivey, J. L., J. Smithers, et al., 2004, "Community Capacity for Adaptation to Climate-Induced Water Shortages: Linking Institutional Complexity and Local Actors." Environmental Management 33(1): 36-47.

55. Ivey, J. L., R. C. de Loë, et al., 2006, "Planning for source water protection in Ontario." Applied Geography 26(3–4): 192-209.

56. Ivey, J. L., R. de Loë, et al., 2006, "An institutional perspective on local capacity for source water protection." Geoforum 37(6): 944-957.

57. Jackson, S., P.-L. Tan, et al., 2012, "Tools to enhance public participation and confidence in the development of the Howard East aquifer water plan, Northern Territory." Journal of Hydrology 474(0): 22-28.

58. Janssen, J. E. B., A. Hoekstra, et al., 2009, "Delineating the Model-Stakeholder Gap: Framing Perceptions to Analyse the Information Requirement in River Management." Water Resources Management 23(7): 1423-1445.

59. Kallis, G., M. Kiparsky, et al., 2009, "Collaborative governance and adaptive management: Lessons from California's CALFED Water Program." Environmental Science & Policy 12(6): 631-643.

60. Karageorgis, A. P., M. S. Skourtos, et al., 2005, "An integrated approach to watershed management within the DPSIR framework: Axios River catchment and Thermaikos Gulf." Regional Environmental Change 5(2-3): 138-160.

61. König, H., M. Sghaier, et al., 2012, "Participatory Impact Assessment of Soil and Water Conservation Scenarios in Oum Zessar Watershed, Tunisia." Environmental Management 50(1): 153-165.

62. Kristensen, S., B. Noble, et al., 2013, "Capacity for Watershed Cumulative Effects Assessment and Management: Lessons from the Lower Fraser River Basin, Canada." Environmental Management: 1-14.

63. Larson, K. L., A. Wiek, et al., 2013, "A comprehensive sustainability appraisal of water governance in Phoenix, AZ." Journal of Environmental Management 116(0): 58-71.

64. Lee, S. and G.-W. Choi, 2012, "Governance in a River Restoration Project in South Korea: The Case of Incheon." Water Resources Management 26(5): 1165-1182.

65. Lemos, M. C. and J. L. F. de Oliveira, 2004, "Can Water Reform Survive Politics? Institutional Change and River Basin Management in Ceará, Northeast Brazil." World Development 32(12): 2121-2137.

66. Lescot, J.-M., P. Bordenave, et al., 2013, "A spatially-distributed cost-effectiveness analysis framework for controlling water pollution." Environmental Modelling & Software 41(0): 107-122.

67. Lienert, J., F. Schnetzer, et al., 2013, "Stakeholder analysis combined with social network analysis provides fine-grained insights into water infrastructure planning processes." Journal of Environmental Management 125(0): 134-148.

68. MacGillivray, B. H. and S. J. T. Pollard, 2008, "What can water utilities do to improve risk management within their business functions? An improved tool and application of process benchmarking." Environment International 34(8): 1120-1131.

69. Mackenzie, J., P.-L. Tan, et al., 2012, "The value and limitations of Participatory Action Research methodology." Journal of Hydrology 474(0): 11-21.

70. Memon, P. A., N. A. Kirk, et al., 2011, "Limits to Ecological Modernisation as a framework for sustainable fresh water governance." Land Use Policy 28(3): 534-541.

71. Menció, A., A. Folch, et al., 2010, "Analyzing Hydrological Sustainability Through Water Balance." Environmental Management 45(5): 1175-1190.

72. Miranda, M., C. Dieperink, et al., 2007, "VOLUNTARY AGREEMENTS IN WATERSHED PROTECTION EXPERIENCES FROM COSTA RICA." Environment, Development and Sustainability 9(1): 1-19.

73. Molina, J. L., J. Bromley, et al., 2010, "Integrated water resources management of overexploited hydrogeological systems using Object-Oriented Bayesian Networks." Environmental Modelling & Software 25(4): 383-397.

74. Molina, J.-L., J. García-Aróstegui, et al., 2011, "Integrated Assessment of the European WFD Implementation in Extremely Overexploited Aquifers Through Participatory Modelling." Water Resources Management 25(13): 3343-3370.

75. Morais, D. C. and A. T. de Almeida, 2012, "Group decision making on water resources based on analysis of individual rankings." Omega 40(1): 42-52.

76. Mouratiadou, I. and D. Moran, 2007, "Mapping public participation in the Water Framework Directive: A case study of the Pinios River Basin, Greece." Ecological Economics 62(1): 66-76.

77. Mutikanga, H., S. Sharma, et al., 2011, "Multi-criteria Decision Analysis: A Strategic Planning Tool for Water Loss Management." Water Resources Management 25(14): 3947-3969.

78. Nesheim, I., D. McNeill, et al., 2010, "The challenge and status of IWRM in four river basins in Europe and Asia." Irrigation and Drainage Systems 24(3-4): 205-221.

79. Nikitina, E., E. Ostrovskaya, et al., 2010, "Towards better water governance in river basins: some lessons learned from the Volga." Regional Environmental Change 10(4): 285-297.

80. Norman, E., G. Dunn, et al., 2013, "Water Security Assessment: Integrating Governance and Freshwater Indicators." Water Resources Management 27(2): 535-551.

81. Nunneri, C. and J. Hofmann, 2005, "A participatory approach for Integrated River Basin Management in the Elbe catchment." Estuarine, Coastal and Shelf Science 62(3): 521-537.

82. Orr, P., J. Colvin, et al., 2007, "Involving stakeholders in integrated river basin planning in England and Wales." Water Resources Management 21(1): 331-349.

83. Osumanu, I., 2008, "Private Sector Participation in Urban Water and Sanitation Provision in Ghana: Experiences from the Tamale Metropolitan Area (TMA)." Environmental Management 42(1): 102-110.

84. Packialakshmi, S., N. K Ambujam, et al., 2011, "Groundwater market and its implications on water resources and agriculture in the southern peri-urban interface, Chennai, India." Environment, Development and Sustainability 13(2): 423-438.

85. Palme, U. and A.-M. Tillman, 2008, "Sustainable development indicators: how are they used in Swedish water utilities?" Journal of Cleaner Production 16(13): 1346-1357.

86. Plummer, R., D. de Grosbois, et al., 2013, "An integrative assessment of water vulnerability in First Nation communities in Southern Ontario, Canada." Global Environmental Change 23(4): 749-763.

87. Pollard, S. and D. Toit, 2011, "Towards Adaptive Integrated Water Resources Management in Southern Africa: The Role of Self-organisation and Multi-scale Feedbacks for Learning and Responsiveness in the Letaba and Crocodile Catchments." Water Resources Management 25(15): 4019-4035.

88. Rodgers, C., N. Giesen, et al., 2007, The GLOWA Volta Project: A framework for water resources decision-making and scientific capacity building in a transnational West African basin. Water Resource Management. (21)295-313.

89. Rohde, S., M. Hostmann, et al., 2006, "Room for rivers: An integrative search strategy for floodplain restoration." Landscape and Urban Planning 78(1–2): 50-70.

90. Roncoli, C., P. Kirshen, et al., 2009, "From Management to Negotiation: Technical and Institutional Innovations for Integrated Water Resource Management in the Upper Comoé River Basin, Burkina Faso." Environmental Management 44(4): 695-711.

91. Silva, V. S., D. Morais, et al., 2010, "A Multicriteria Group Decision Model to Support Watershed Committees in Brazil." Water Resources Management 24(14): 4075-4091.

92. Simon, U., R. Brüggemann, et al., 2004, "Aspects of decision support in water management—example Berlin and Potsdam (Germany) I—spatially differentiated evaluation." Water Research 38(7): 1809-1816.

93. Sudha, M. C., S. Ravichandran, et al., 2013, "Water Bodies Protection Index for assessing the sustainability status of lakes under the influence of urbanization: a case study of south Chennai, India." Environment, Development and Sustainability: 1-15.

94. Swatuk, L. A. and P. K. Kgomotso, 2007, "The challenges of supplying water to small, scattered communities in the Lower Okavango Basin (LOB), Ngamiland, Botswana: An evaluation of government policy and performance." Physics and Chemistry of the Earth, Parts A/B/C 32(15–18): 1264-1274.

95. Tan, P.-L., C. Baldwin, et al., 2012, "Water planning in the Condamine Alluvium, Queensland: Sharing information and eliciting views in a context of overallocation." Journal of Hydrology 474(0): 38-46.

96. Tapela, B. N., 2002, "The challenge of integration in the implementation of Zimbabwe's new water policy: case study of the catchment level institutions surrounding the Pungwe–Mutare water supply project." Physics and Chemistry of the Earth, Parts A/B/C 27(11–22): 993-1004.

97. Tidwell, V., H. Passell, et al., 2004, "System dynamics modeling for community-based water planning: Application to the Middle Rio Grande." Aquatic Sciences 66(4): 357-372.

98. Tippett, J., 2005, "The value of combining a systems view of sustainability with a participatory protocol for ecologically informed design in river basins." Environmental Modelling & Software 20(2): 119-139.

99. Tjandraatmadja, G., A. K. Sharma, et al., 2013, "A Decision Support Methodology for Integrated Urban Water Management in Remote Settlements." Water Resources Management 27(2): 433-449.

100. Van Cauwenbergh, N., D. Pinte, et al., 2008, "Multi-objective, multiple participant decision support for water management in the Andarax catchment, Almeria." Environmental Geology 54(3): 479-489.

101. Vedwan, N., S. Ahmad, et al., 2008, "Institutional Evolution in Lake Okeechobee Management in Florida: Characteristics, Impacts, and Limitations." Water Resources Management 22(6): 699-718.

102. Vliet, M. and K. Kok, 2013, "Combining backcasting and exploratory scenarios to develop robust water strategies in face of uncertain futures." Mitigation and Adaptation Strategies for Global Change: 1-32.

103. Werff, P., 2004, "Stakeholder responses to future flood management ideas in the Rhine River Basin: nature or neighbour in Hell's Angel." Regional Environmental Change 4(2-3): 145-158.

104. Xenarios, S. and I. Tziritis, 2007, "Improving pluralism in Multi Criteria Decision Aid approach through Focus Group technique and Content Analysis." Ecological Economics 62(3–4): 692-703.

105. Xenarios, S. and K. Bithas, 2009, "Welfare Improvisation from the Receiving Waters of Urban Wastewater Systems in the Context of the Water Framework Directive." Water Resources Management 23(5): 981-1000.

106. Zeng, Y., Y. Cai, et al., 2012, "Development of a web-based decision support system for supporting integrated water resources management in Daegu city, South Korea." Expert Systems with Applications 39(11): 10091-10102.

107. Acreman, M., 2005, "Linking science and decision-making: features and experience from environmental river flow setting." Environmental Modelling & Software 20(2): 99-109.

108. Albrecht, J., 2013, "The Europeanization of water law by the Water Framework Directive: A second chance for water planning in Germany." Land Use Policy 30(1): 381-391.

109. Antunes, P., G. Kallis, et al., 2009, "Participation and evaluation for sustainable river basin governance." Ecological Economics 68(4): 931-939.

110. Brown, C., 2003, "New directions in binational water resource management in the U.S.-Mexico borderlands." The Social Science Journal 40(4): 555-572.

111. Brown, R., 2008, "Local Institutional Development and Organizational Change for Advancing Sustainable Urban Water Futures." Environmental Management 41(2): 221-233.

112. Brown, R., R. Ashley, et al., 2011, "Political and Professional Agency Entrapment: An Agenda for Urban Water Research." Water Resources Management 25(15): 4037-4050.

113. Brugge, R. and J. Rotmans, 2007, "Towards transition management of European water resources." Water Resources Management 21(1): 249-267.

114. Burgin, S., T. Webb, et al., 2013, "Stakeholder engagement in water policy: Lessons from peri-urban irrigation." Land Use Policy 31(0): 650-659.

115. Carter, N., R. D. Kreutzwiser, et al., 2005, "Closing the circle: linking land use planning and water management at the local level." Land Use Policy 22(2): 115-127.

116. Cashman, A., 2006, "Water regulation and sustainability 1997–2001: Adoption or adaptation?" Geoforum 37(4): 488-504.

117. Chikozho, C., 2008, "Globalizing Integrated Water Resources Management: A Complicated Option in Southern Africa." Water Resources Management 22(9): 1241-1257.

118. Cowie, G. M. and S. R. Borrett, 2005, "Institutional perspectives on participation and information in water management." Environmental Modelling & Software 20(4): 469-483.

119. Croke, B. F. W., J. L. Ticehurst, et al., 2007, Integrated assessment of water resources: Australian experiences. Integrated Assessment of Water Resources and Global Change. E. Craswell, M. Bonnell, D. Bossio, S. Demuth and N. Giesen, Springer Netherlands: 351-373.

120. De Stefano, L., 2010, "Facing the water framework directive challenges: A baseline of stakeholder participation in the European Union." Journal of Environmental Management 91(6): 1332-1340.

121. De Stefano, L., J. Pedraza Gilsanz, et al., 2010, "A Methodology for the Evaluation of Water Policies in European Countries." Environmental Management 45(6): 1363-1377.

122. deKok, J.-L. and H. G. Wind, 2003, "Design and application of decision-support systems for integrated water management: lessons to be learnt." Physics and Chemistry of the Earth, Parts A/B/C 28(14–15): 571-578.

123. Donnellan, A., 2007, "Is the National Water Commission an appropriate institutional and governance structure to address the fundamental requirements of the water and irrigation industry in Australia?" The Environmentalist 27(1): 13-23.

124. Faby, J.-A., G. Neveu, et al., 2005, "Towards a European-wide exchange network for improving dissemination of Integrated Water Resources Management Research outcomes." Environmental Science & Policy 8(3): 307-319.

125. Fraternali, P., A. Castelletti, et al., 2012, "Putting humans in the loop: Social computing for Water Resources Management." Environmental Modelling & Software 37(0): 68-77.

126. Fuller, B. W., 2009, "Surprising cooperation despite apparently irreconcilable differences: Agricultural water use efficiency and CALFED." Environmental Science & Policy 12(6): 663-673.

127. Gearey, M. and P. Jeffrey, 2006, "Concepts of legitimacy within the context of adaptive water management strategies." Ecological Economics 60(1): 129-137.

128. Gourbesville, P., 2008, "Integrated river basin management, ICT and DSS: Challenges and needs." Physics and Chemistry of the Earth, Parts A/B/C 33(5): 312-321.

129. Hall, D. and E. Lobina, 2007, "International actors and multinational water company strategies in Europe, 1990–2003." Utilities Policy 15(2): 64-77.

130. Harrison, P. A., P. M. Berry, et al., 2008, "Impacts of socio-economic and climate change scenarios on wetlands: linking water resource and biodiversity meta-models." Climatic Change 90(1-2): 113-139.

131. Harvey, B. and A. Schaefer, 2001, "Managing Relationships with Environmental Stakeholders: A Study of U.K. Water and Electricity Utilities." Journal of Business Ethics 30(3): 243-260.

132. Heinz, I., M. Pulido-Velazquez, et al., 2007, "Hydro-economic Modeling in River Basin Management: Implications and Applications for the European Water Framework Directive." Water Resources Management 21(7): 1103-1125.

133. Holt, L., 2005, "Utility service quality—Telecommunications, electricity, water." Utilities Policy 13(3): 189-200.

134. Horlitz, T., 2007, "The Role of Model Interfaces for Participation in Water Management." Water Resources Management 21(7): 1091-1102.

135. Iglesias, A., L. Garrote, et al., 2007, "Challenges to Manage the Risk of Water Scarcity and Climate Change in the Mediterranean." Water Resources Management 21(5): 775-788.

136. Ison, R., N. Röling, et al., 2007, "Challenges to science and society in the sustainable management and use of water: investigating the role of social learning." Environmental Science & Policy 10(6): 499-511.

137. Koehler, B. and T. Koontz, 2008, "Citizen Participation in Collaborative Watershed Partnerships." Environmental Management 41(2): 143-154.

138. Macleod, C. J. A., D. Scholefield, et al., 2007, "Integration for sustainable catchment management." Science of The Total Environment 373(2–3): 591-602.

139. Mirchi, A., K. Madani, et al., 2012, "Synthesis of System Dynamics Tools for Holistic Conceptualization of Water Resources Problems." Water Resources Management 26(9): 2421-2442.

140. Mooney, C., C. Baldwin, et al., 2012, "Transparency and trade-offs in water planning." Journal of Hydrology 474(0): 66-73.

141. Moraes, A. F. J. d. and C. Rocha, 2013, "Gendered waters: the participation of women in the 'One Million Cisterns' rainwater harvesting program in the Brazilian Semi-Arid region." Journal of Cleaner Production(0).

142. Morais, D. C. and A. T. de Almeida, 2007, "Group decision-making for leakage management strategy of water network." Resources, Conservation and Recycling 52(2): 441-459.

143. Mugabi, J., S. Kayaga, et al., 2007, "Strategic planning for water utilities in developing countries." Utilities Policy 15(1): 1-8.

144. Nel, E. L., T. R. Hill, et al., 2007, "Multi-Stakeholder Driven Local Economic Development: Reflections on the Experience of Richards Bay and the uMhlathuze Municipality." Urban Forum 18(2): 31-47.

145. Nguyen, T. G., J. L. de Kok, et al., 2007, "A new approach to testing an integrated water systems model using qualitative scenarios." Environmental Modelling & Software 22(11): 1557-1571.

146. Pahl-Wostl, C., G. Holtz, et al., 2010, "Analyzing complex water governance regimes: the Management and Transition Framework." Environmental Science & Policy 13(7): 571-581.

147. Pearson, L., A. Coggan, et al., 2010, "A Sustainable Decision Support Framework for Urban Water Management." Water Resources Management 24(2): 363-376.

148. Qadir, M., B. R. Sharma, et al., 2007, "Non-conventional water resources and opportunities for water augmentation to achieve food security in water scarce countries." Agricultural Water Management 87(1): 2-22.

149. Reichert, P., M. Borsuk, et al., 2007, "Concepts of decision support for river rehabilitation." Environmental Modelling & Software 22(2): 188-201.

150. Ridder, D. and C. Pahl-Wostl, 2005, "Participatory Integrated Assessment in local level planning." Regional Environmental Change 5(4): 188-196.

151. Rijsberman, M. A. and F. H. M. van de Ven, 2000, "Different approaches to assessment of design and management of sustainable urban water systems." Environmental Impact Assessment Review 20(3): 333-345.

152. Roach, K., 2013, "Texas water wars: how politics and scientific uncertainty influence environmental flow decision-making in the Lone Star state." Biodiversity and Conservation 22(3): 545-565.

153. Saravanan, V. S., 2008, "A systems approach to unravel complex water management institutions." Ecological Complexity 5(3): 202-215.

154. Singh, A., C. Singai, et al., 2009, "Inclusive Water Governance: a Global Necessity. Lessons from India." Transition Studies Review 16(2): 598-608.

155. Smith, L. D. and K. Porter, 2010, "Management of catchments for the protection of water resources: drawing on the New York City watershed experience." Regional Environmental Change 10(4): 311-326.

156. Srdjevic, B., 2007, "Linking analytic hierarchy process and social choice methods to support group decision-making in water management." Decision Support Systems 42(4): 2261-2273.

157. Stakhiv, E. and B. Stewart, 2010, "Needs for Climate Information in Support of Decision-Making in the Water Sector." Procedia Environmental Sciences 1(0): 102-119.

158. Starkl, M. and N. Brunner, 2004, "Feasibility versus sustainability in urban water management." Journal of Environmental Management 71(3): 245-260.

159. Tippett, J., B. Searle, et al., 2005, "Social learning in public participation in river basin management—early findings from HarmoniCOP European case studies." Environmental Science & Policy 8(3): 287-299.

160. Uitto, J. I., 2004, "Multi-country cooperation around shared waters: role of monitoring and evaluation." Global Environmental Change 14, Supplement(0): 5-14.

161. Webb, T., S. Burgin, et al., 2009, "Action Research for Sustainable Water Futures in Western Sydney: Reaching Beyond Traditional Stakeholder Engagement to Understand Community Stakeholder Language and Its Implications for Action." Systemic Practice and Action Research 22(1): 1-14.

162. Wiek, A. and K. Larson, 2012, "Water, People, and Sustainability—A Systems Framework for Analyzing and Assessing Water Governance Regimes." Water Resources Management 26(11): 3153-3171.

163. Wilder, M. and P. Romero Lankao, 2006, "Paradoxes of Decentralization: Water Reform and Social Implications in Mexico." World Development 34(11): 1977-1995.

164. Wilson, R. T., 2007, "Perceptions, practices, principles and policies in provision of livestock water in Africa." Agricultural Water Management 90(1–2): 1-12.

165. Winz, I., G. Brierley, et al., 2009, "The Use of System Dynamics Simulation in Water Resources Management." Water Resources Management 23(7): 1301-1323.

166. Yamout, G. and D. Jamali, 2007, "A critical assessment of a proposed public private partnership (PPP) for the management of water services in Lebanon." Water Resources Management 21(3): 611-634.

Chapter 5

Proposing a Secure and Reliable System for Critical Pipeline Infrastructure Based on Wireless Sensor Network

Fawaz Alsaade

Collage of computer Sciences and Information Technology, King Faisal University, Al-Ahsa, Saudi Arabia

ABSTRACT

A combined approach to enhance the security of oil and water supply pipeline infrastructures is proposed and investigated. The most important factor affecting the performance of the traditional system is the mannual patrolling which is difficult and only provides observation where the patrolling team is present. This study highlighted that by deploying a combination of wireless sensor network in conjunction with the conventional trends and microwave network, the time of reporting any leakage to the control room can be reduced considerably. This in turn provides ability to protect the oil pipelines from further loss or damage and discontinuation of operation. The simulation tests were carried under three different scenarios i.e., minor level leakage, major level leakage and any criminal activity to break a pipeline. Based on the simulation results, it is demonstrated that the reliability of monitoring oil pipelines provided by the new proposed system can be increased. The paper presents the motivation for and the potential advantages of the proposed WSN system for enhancing the security of oil and water supply pipelines.

KEYWORDS

Wireless sensor network, security, oil water, critical pipeline infrastructure, reliability, microwave network

1. INTRODUCTION

In Saudi Arabia, oil and gas industries heavily depend on more than 152000 km of oil pipelines installed across the country. This network of pipelines connects the shipping ports, refineries and oil/gas wells. Moreover, Saudi Arabia is now considered the world's largest producer of desalinated water supplying drinking water to major urban and industrial areas through a network of water pipelines over more than 4,800 km distance. For example, the capital city of Riyadh with homes more than a million completely depends on the water transferred through huge and long pipelines from Al Shoaiba Desalination Plant in Al-Jubail, Eastern part of Saudi Arabia. Currently, the main concern of Saudi government is to ensure safe and smooth transfer of oil, gas and water to consumers using modern up-to-date technology and tools. But still, there is a possibility of internal damage to pipelines which may be caused due to any terrorist activities, theft or any natural calamities.

Presently, Saudi ARAMCO is the largest company for oil exploring and transporting to the ports. It also manages and controls the pipeline infrastructures by monitoring the pressure, maintaining constant level of pressure at different points and also dividing oil pipelines into different segments. The company still uses the physical patrolling with Ardmco employed teams to monitor its safety. It is believed that a new era of technologies like wireless sensor network can play an important role in improving the process of monitoring such important pipelines.

Sensor network is becoming an increasingly important requirement in a variety of applications such as target detection, surveillance of enemy activities in a battle field, health, chemical use and counter terrorism (Chong and Kumar, 2003; Gutierrez et al., 2003; De et al., 2002). Some other examples are environment and habitat monitoring, home automation, traffic control, etc. Another possible example is using wireless sensor networks for protecting and monitoring pipeline infrastructure systems especially for oil, gas and water huge pipeline infrastructures (Carrillo et al., 2002; Ellul, 1989; Awawdeh et al., 2006; Suheil et al., 2007).

This study proposes a method to improve the existing oil pipeline security and monitoring system which is cost effective, ensure quality, highly efficient and reliable. This can be achieved by incorporating Wireless Sensor Network (WSN) based system with the existing system. Because WSN sensor nodes are capable to measure a given physical environment in greater detail. A wireless sensor network can be described as a collection of sensor nodes which co-ordinate to perform some specific actions. Unlike conventional networks, sensor networks depend on dense deployment and co-ordination to carry out their tasks. These unique characteristics make sensor network more beneficial and ease to use over the traditional network.

Many investigators concluded that Wireless Sensor Network (WSN) in the oil and gas industry shed some light on a model of tracking flow-induced vibration to provide means of detection and early warning of integrity loss in pipelines network (Ellul, 1989; Gerasimov and Simon, 2002; Hadim et al., 2006; Hadim and Mohamed, 2006; Jawhar and Wu, 2005; Liao et al., 2002; Nelakuditi et al., 2002; Stoianov et al., 2007). Recently, Suheil et al. (2007) proposed a model for pipeline monitoring but the technical information is limited only. Changsoo et al. (2007) proposed a method for calculating number of sinks and sensors required for pipeline monitoring. Umeadi and Jones (2005) reported some findings of a laboratory based test program to evaluate the potential for vibration sound emission detection in pipelines integrity monitoring. Stoianov et al. (2007) presented a system called PIPENET for monitoring large diameter bulk-water transmission pipelines. The above review concluded that in public domain literature limited information are available for WSN based pipelines security, condition and information monitoring.

Akan and Akyildiz (2005) presented an even-to-sink transport protocol for reliable transport in wireless sensor network. However, it should be noted that due to the unique arrangements of the wireless sensors in the case of pipelines monitoring none of the above mentioned models can be adopted for such a case. Based on the above review it may be observed that the topic of continuous real time security and informative reliability of wireless sensor network for pipelines monitoring has not been approached. The proposed system, in this study, focuses on the continuous monitoring security and real time information for the critical pipeline infrastructure.

2. PROPOSED APPROACH

As stated earlier that the oil transportation across the Kingdom and the rest of the world, is mainly based on the huge critical pipeline infrastructure. These pipelines are mostly laid down in straight and multi curve status (Fig. 1). These critical pipelines are mostly lay down underground with little depth for their protection. The surveillance and security for those pipelines is done through manual patrolling and monitoring.

Manual patrolling is hard, difficult and only provides observations where the patrolling team is present. It is believed that a technique based on joining wireless sensor network with the conventional trends and microwave network together could improve the reliability and quality of monitoring of the oil pipeline infrastructure. The proposed system would detect, identify and localize major anomalies such as theft, terrorism, damage or any disturbance of the pipeline infrastructure at run time. This is achieved because the pipeline infrastructure was divided into different segments with each having a length of 10 km.

The sensor nodes were placed closed to the buried pipes in each segment. Those sensor nodes have coverage area around each node which can be considered all most same in all directions theoretically as shown in Fig. 2. The circle shows the coverage area for each sensor node. Sensors can be closely deployed to the straight and multi direction pipelines as shown in Fig. 3.

The general architecture of the proposed system is shown in Fig. 4. The proposed system consists of sensor nodes, sink nodes, base station and a control room which might be located at head office. Each component of the system perform certain tasks.

For example, the wireless sensor nodes timely monitor the oil pipelines and report to the sink node if any leakage, breakage or any theft activity occurs. The sink node transfers the collected data to the base station (tower with microwave). Then, base station transfers such data along with its current calculated location to the satellite. A central control room receives all such data from the satellite. The proposed system covers all these processes in few seconds with its exact information of the problem site. Therefore, it is believed that this mechanism would boost the security at the run time and improve the efficiency and reliability of the proposed oil transferring system.

Figure 1: Oil refinery with laying pipeline structure

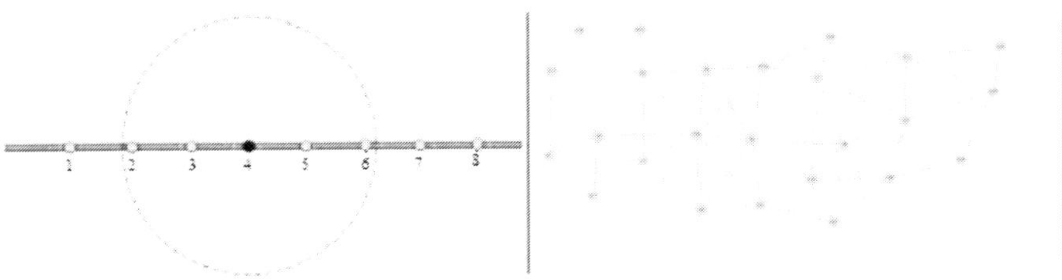

Figure 2: Working of WSN nodes with a segment of pipeline

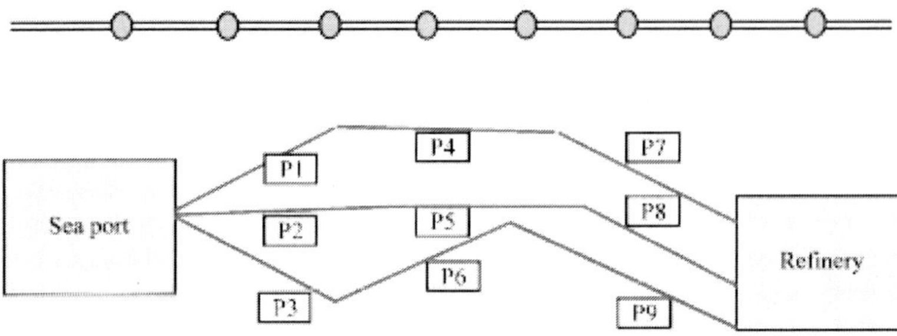

Figure 3. Simple pipeline structure for oil fields

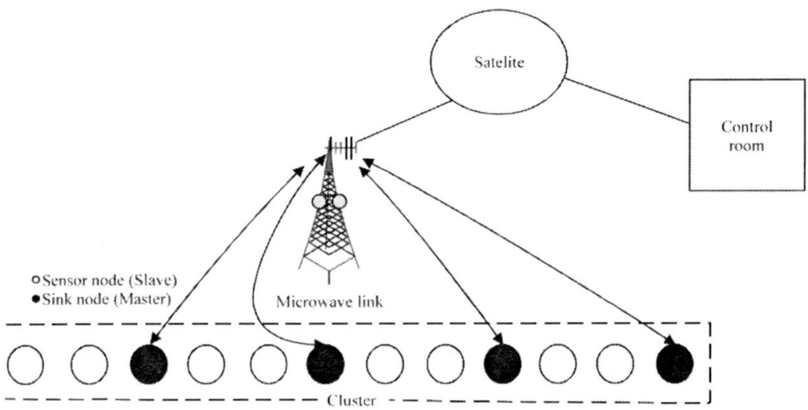

Figure 4: A proposed system architecture

3. SIMULATIONS

The proposed system was tested through different simulation tests and their results were compared to the conventional available security system. Network Simulator 2 (NS-2) is used for simulation purpose. Such simulator provides a graphical interface called Network Animator (NAM). This graphical interface helps to generate trace files in order to control the simulations during run time. An example of the animated tool NAM is shown in Fig. 5.

Different parameters were considered during the simulation process. Some of those parameters are shown in Table 1.

4. RESULTS

As stated earlier that oil pipelines are mostly lay down underground at shallow depth for their protection. Therefore the process of noting the leakage at limited time seems to be difficult when using the existing system. This is because security of pipeline infrastructures in such system normally checked and monitored with patrolling and pressure measuring system from one to another end. An important requirement for the effective operation of the proposed system is the capability for this system to minimize the time of reporting any leakage to the control room. This in turn can provide ability to protect the oil pipelines from further loss or damage and discontinuation of operation. Various simulations tests were applied to check the security, reliability and real time monitoring for the proposed system and then compared it with the existing system approaches. These simulation tests were conducted under three different assumptions. Each assumption is evaluated based on 11 different events. These events present different locations.

The following subsections “minor level of leakage” to “criminal activity to break a pipeline” introduce those assumptions as well as present the experimental investigations and an analysis of the results. The results are discussed based on three different parameters. These are Event occurrence Data Time (EDT), Access Time of Existing System (ATES) and Access Time of Proposed System (ATPS).

Minor level leakage: In this assumption, different simulation testes were applied to track the minor level leakage in the pipeline and then reported to the control room. This includes the security, reliability and monitoring access time for both systems. Figure 6 clearly shows the vast difference in reporting or access time between the conventional and the proposed system.

It can be observed from Fig. 6 that the conventional system reported the event in about 330 sec after its occurrence time. On the other hand, the proposed system can reduce the access time to only 15 sec after event occurrence. The reduction in access time achieved with such system is more than 10 times efficient as compared to the conventional system. These results are in agreement with the earlier suggestions that the use of conventional system require more time to monitor this type of leakage.

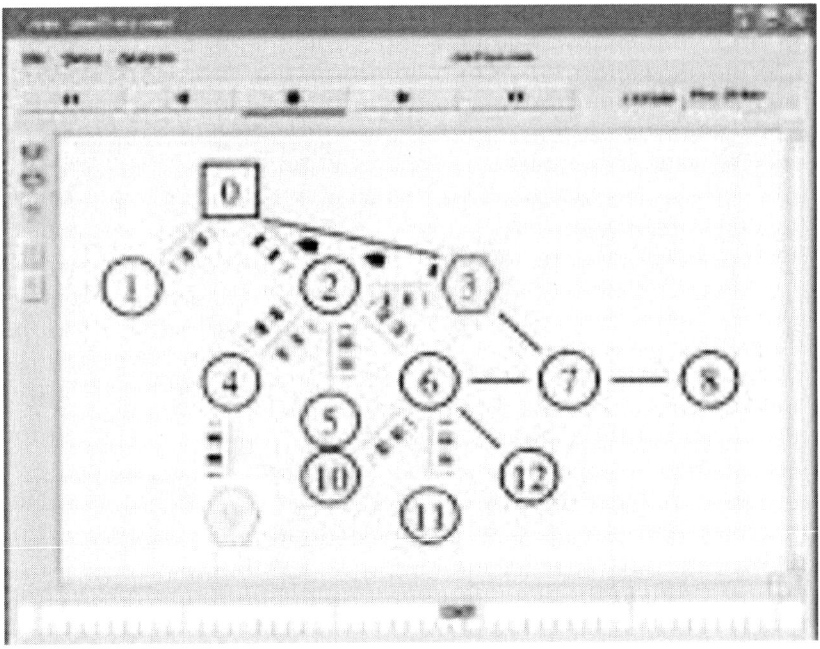

Figure 5: NS-2 NAM graphic animator tool

The possible reason is that this system can only monitor the leakage loss through its pressure observation techniques for different pipeline segments. It even required more delay in case of nights patrolling, while the proposed system can work with same quality and efficiency all the time.

Major level leakage: In this assumption, simulations were applied to monitor the major level leakage. This level of leakage may be caused by major damage or pipeline breakage due to pressure. It can be seen in Fig. 7 that the conventional system leads to an access time which is slightly better than the one offered by the same system in the previous section. The reason for such behaviour is that the conventional system could report this level of leakage quickly due to the drop-down of the pressure which can be noticed at different control segments. Therefore, this level of leakage can easily be monitored during night timings, as it becomes a sudden cause of drop in pressure and can easily be monitored through pressure gauges. However, the performance of such system, in terms of access time, is still well below that of the proposed system. The use of proposed system is seen to reduce the access time by about 7%. This in turn can save major resources and enhance their safety and security.

Table 1: Simulation parameters

Parameter	Values
Network area	200*200 m or 400*400 m
Number of sensor	100 or 400
Location of sink	Uniform random
Radio range	Center of area
MAC layer	IEEE 802.11
Unusual event sources	4
Routine data sources probability	p
Failure rate	f
Time-out constant ξ	1/r
Delay for retransmission M	0.02 s
Data rate of unusual events	λ_U
Data rate of routine data	λ_R

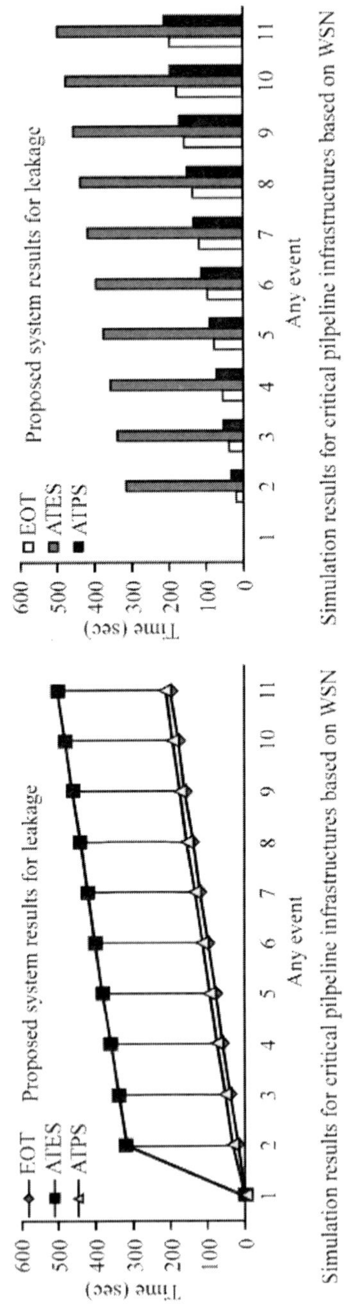

Figure 6: Performance of proposed and conventional systems based on minor leakage monitoring in oil pipelines

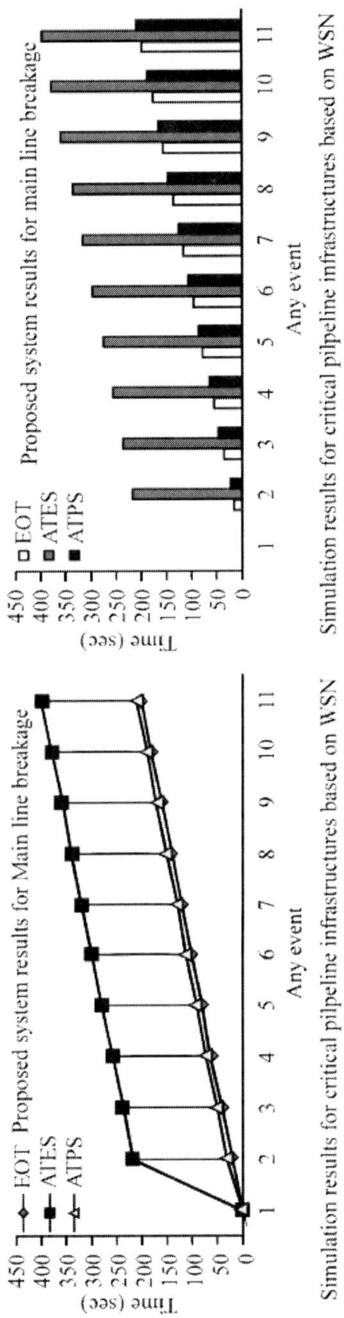

Figure 7: Performance of proposed and conventional systems based on major leakage monitoring in oil pipelines

Criminal activity to break a pipeline: In this assumption, simulations were applied to monitor the status of oil pipelines and prevent any sabotage operation. It is observed from Fig. 8 that the lowest access time is obtained when monitoring of the pipelines is based on the use of the proposed system. The access time, obtained by the proposed system, is 5 times faster than the one obtained by conventional system. In such scenario, the proposed system can provide a report even before the sabotage activity occurred.

This is, as explained earlier, because sensor nodes will be deployed across the pipelines to monitor any unwanted event occurrence. On the other hand, conventional system can only report an event when it occurred since it is based on patrolling and pressure measuring system from one to another end.

A performance assessment of the overall results in terms of access time is presented in Fig. 9.

It was observed from Fig. 9 that using the proposed system in monitoring oil pipelines appears to provide better performance in terms of reducing the access time. These outcomes confirm the earlier suggestion that the reliability of monitoring oil pipelines can be further increased if a mechanism of joined wireless sensor network with the conventional trends and microwave network is applied.

Another major benefit offered by such combined system is the capability to monitor the critical pipeline infrastructure at run time under any hard circumstances which is a leading edge over the conventional security system. The findings of the study agree with many researchers. For example: Cinque et al. (2006) addressed the current limitations and posed the reliability requirements for dynamic structure monitoring using wireless sensor network. Abo El-Fotoh et al. (2006) approached the problem of modeling and evaluate the coverage oriented probability of WSN subject to common cause failure. Xing and Michel (2006) defined a WSN reliability measure considering the aggregate flow of sensor data into a sink node. Chiang et al. (2004) dealt with the problem of reliability and security modeling in an integrated manner. Cai et al. (2006) proposed and evaluated scalable architecture of WSN nodes for increased availability. Shrestha et al. (2007) shed some light on the problem of reliability modeling for large scale wireless sensor network.

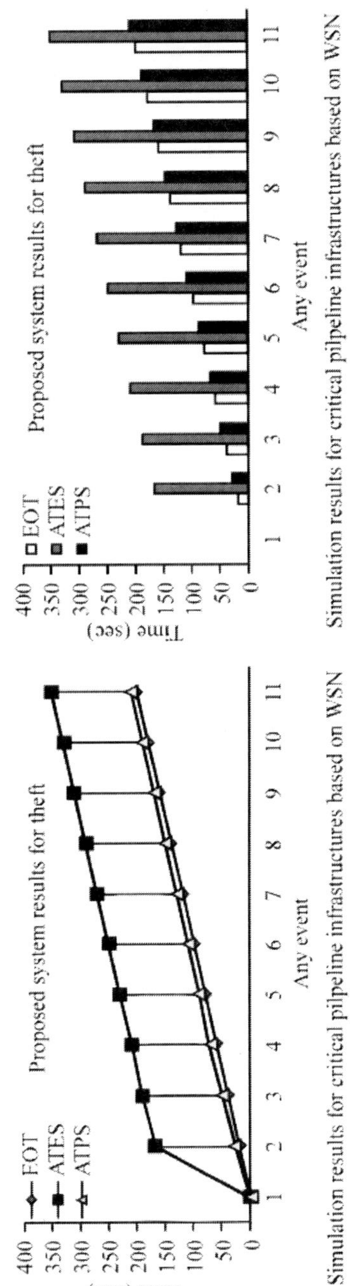

Figure 8: Effectiveness of proposed system in monitoring the critical oil pipelines from criminal activities

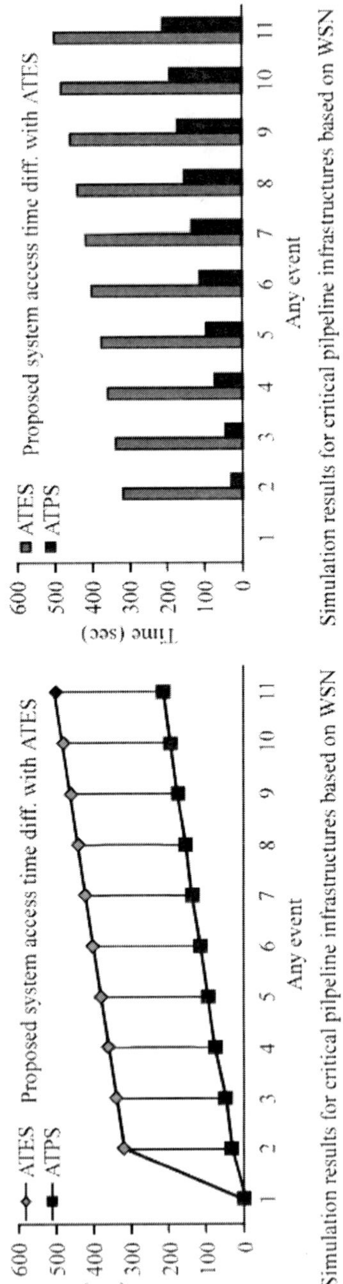

Figure 9: Overall performance of the proposed and existing monitoring oil systems

5. CONCLUSION

A study of the use of wireless sensor network in conjunction with the conventional trends and microwave network, in monitoring oil pipelines, is presented. Simulation tests were carried under three different scenarios to check the security, reliability and real time monitoring for the proposed system and then compare it with existing system approaches. For all scenarios, the results showed that the use of the proposed method leads to better performance, in terms of access time, than the conventional system. This in turn increases the reliability of monitoring oil pipelines and provides ability to protect the oil pipelines from further loss or damage and discontinuation of operation. Additionally, it has been shown that the proposed system is capable to monitor the critical pipeline infrastructure at run time under any hard circumstances.

REFERENCES

1. Abo El-Fotoh, H.M.F, E.S. El-Mallah and H. Hassanein, 2006. Sm on the reliability of wireless sensor networks. Proceedings of the International Conference on Communications, June 11-15, 2006, IEEE., pp: 3455-3460.

2. Akan, O.B. and I.F. Akyildiz, 2005. Event-to-sink reliable transport in wireless sensor networks. IEEE/ACM Trans. Networking, 13: 1003-1016.

3. Awawdeh, A., S.T.S. Bukkapatnam, S.R.T. Kumara, C. Bunting and R. Komanduri, 2006. Wireless sensing of flown-induced vibrations for pipeline integrity monitoring. Proceedings of the Workshop on Sensor Array and Multichannel Signal Processing, July 12-14, 2006, IEEE., China, pp: 114-117.

4. Cai, W., J. Xinyu, Y. Zhang, C. Kangsherg and J. Tang, 2006. Research on reliability model of large scale wireless sensor networks. Proceedings of the International Conference on Wireless Communications, Networking and Mobile Computing, September 22-24, 2006, Wuhan, China, pp: 1-4.

5. Carrillo, A., E. Gonzalez, A. Rosas and A. Marquez, 2002. New distributed optical sensor for detection and localization of liquid leaks. Pat I: Experimental studies. Sensor Actuators, 99: 229-235.

6. Changsoo, O.K., H. Thadakarmalla, U. Raghavan, S. Kumara, S.G. Kim, X. Zhang and S. Bukkapatnam, 2007. Optimal transmission power in self sustainable sensor networks for pipeline monitoring. Proceedings of the International Conference on Automation Science and Engineering, May 18, 2007, IEEE., pp: 591-596.

7. Chiang, M.W., Z. Zilic, K. Radecka and J.S. Chenard, 2004. Architectures of increased availability wireless sensor network nodes. Proceedings of the International Test Conference, October 26-28, 2004, Charlotte, NC., USA., pp: 1232-1241.

8. Chong, C.Y. and S.P. Kumar, 2003. Sensor networks: Evolution, opportunities and challenges. Proc. IEEE, 91: 1247-1256.

9. Cinque, M., D. Cotroneo, G. De Caro and M. Pelella, 2006. Reliability requirements of wireless sensor networks for dynamic structural monitoring. Proceedings of the International Workshop on Applied Software Reliability, June 26, 2006, Philadelphia, PA., USA., pp: 8-13.

10. De, S., S.K. Das, H. Wu and C. Qiao, 2002. A resource efficient RTQoS routing protocol for mobile ad hoc networks. Proceedings of the 5th International Symposium on Wireless Personal Multimedia Communications, October 27-30, 2002, Honolulu, HI., USA., pp: 257-261.

11. Ellul, I., 1989. Pipeline leak detection. Chem. Eng., 461: 39-44.

12. Gerasimov, I. and R. Simon, 2002. Performance analysis for ad hoc QoS routing protocols. Proceedings of the International Mobility and Wireless Access Workshop, October 12, 2002, Texas, pp: 87-94.

13. Gutierrez, J.A., E.H. Callaway and R.L. Barrett, 2003. IEEE 802.15.4 Low-Rate Wireless Personal Area Networks: Enabling Wireless Sensor Networks. Institute of Electrical and Electronics Engineering, Piscataway, NJ., ISBN-13: 978-0738135571, Pages: 155.

14. Hadim, S. and N. Mohamed, 2006. Middleware: Middleware challenges and approaches for wireless sensor networks. IEEE Distributed Syst. Online, 7: 1-1.

15. Hadim, S., J. Al-Jaroodi and N. Mohamed, 2006. Trends in middleware for mobile Ad Hoc networks. J. Commun., 1: 11-21.

16. Jawhar, I.J. and J. Wu, 2005. Qos support in TDMA-based mobile ad hoc networks. J. Comput. Sci. Technol., 20: 797-910.

17. Liao, W.H., Y.C. Tseng and K. P. Shih, 2002. A TDMA-based bandwidth reservation protocol for QoS routing in a wireless mobile ad hoc network. Communications, ICC 2002. IEEE Intl. Conf. Autom. Scie. Eng, 5: 3186-3190.

18. Nelakuditi, S., Z.L. Zhang, R.P. Tsang and D. H.C. Du, 2002. Adaptive proportional routing: A localized QoS routing approach. IEEE/ACM Trans. Networking, 10: 790-804.

19. Shrestha, A., L. Xing and H. Liu, 2007. Modeling and evaluating the reliability of wireless sensor networks. Proceedings of the Annual Reliability and Maintainability Symposium, January 22-25, 2007, Orlanndo, Florida, pp: 186-191.

20. Stoianov, I., L. Nachman, S. Madden, T. Tokmouline and C.M. Pipenet, 2007. A wireless sensor network for pipeline monitoring. Proceedings of the 6th Internationl Symposium on Information Processing in Sensor Netwoks, April 25-27, 2007, Cambridge, Massachusetts, USA., pp: 246-273.

21. Suheil, S., H. Nasr and R. Garelli, 2007. Wireless sensors enhance system safety and efficiency. Oil Gas J., 105: 66-68.

22. Umeadi, B.B.N. and K.G Jones, 2005. Integrated wireless sensors in oil pipeline integrity monitoring. Analysis and Report, University of Greenwich UK.

23. Xing, L. and H.E. Michel, 2006. Integrated modeling for wirless sensor netwoks reliability and security. Proceedings of the Annual Reliability and Maintainability Symposium, June 23-26, 2006, IEEE., pp: 594-600.

Chapter 6

Dealing with Uncertainties in Fresh Water Supply: Experiences in the Netherlands

Wil Thissen[1,7] , Jan Kwakkel [1] , Marjolein Mens [2,3] , Jeroen van der Sluijs[4,5] , Sara Stemberger[4] , Arjan Wardekker[4], Diana Wildschut[4,6]

[1]Faculty of Technology, Policy and Management, Delft University of Technology, Jaffalaan 5, 2628 BX Delft, The Netherlands

[2] Department of Flood and Drought Risk Analysis, Deltares, P.O. Box 17, 2600 MH Delft, The Netherlands

[3] Twente Water Centre, Twente University, P.O. Box 217, 7500 AE Enschede, The Netherlands

[4] Centre for the Study of the Sciences and the Humanities and Department of Chemistry, University of Bergen, PO Box 7805, 5020 Bergen, Norway

[5] Copernicus Institute of Sustainable Development, Utrecht University, PO Box 80115, 3508 TC Utrecht, Netherlands

[6] Cooperative University of Amersfoort, Geldersestraat 6, 3812 PP Amersfoort, The Netherlands

[7] UNESCO-IHE Institute for Water Education, Westvest 7, 2611 AX Delft, The Netherlands

ABSTRACT

Developing fresh water supply strategies for the long term needs to take into account the fact that the future is deeply uncertain. Not only the extent of climate change and the extent and nature of its impacts are unknown, also socio-economic conditions may change in unpredictable ways, as well as social preferences. Often, it is not possible to find solid ground for estimating probabilities for the relevant range of imaginable possible future developments. Yet, some of these may have profound impacts and consequences for society which could be reduced by timely proactive adaptation. In response to these and similar challenges, various approaches, methods and techniques have been proposed and are being developed to specifically address long-term strategy development under so-called deep uncertainty. This paper, first, offers a brief overview of developments in the field of planning under (deep) uncertainty. Next, we illustrate application of three different approaches to fresh water provision planning under uncertainty in case studies in the Netherlands: a resilience approach, oriented to (re) designing fresh water systems in such a way that they will be less vulnerable, resp. will be able to recover easily from future disturbances; a robustness approach, oriented to quantitative assessment of system performance for various system configurations (adaptation options) under a range of external disturbances, and an exploratory modeling approach, developed to explore policy effectiveness and system operation under a very wide set of assumptions about future conditions.

KEYWORDS

Fresh water supply, Uncertainties, Adaptation, Resilience Robustness

1. UNCERTAINTIES: NEW CHALLENGES FOR DECISION MAKING ON FRESH WATER SUPPLY SYSTEMS

Decision making and planning for fresh water supply are challenging because of a number of different reasons: First, a variety of individual measures or tactics need to be considered at both the supply and the demand side, including regulation, adaptation of infrastructure elements, subsidies or fines, operational control and coordination, etc., and appropriate combinations of these need to be made in the process of policy design. Second, different scales need to be considered, such as the local, regional, national and river basin scales, each having their associated governmental bodies. Third, a range of sometimes conflicting interests needs to be considered. Fourth, various uncertainties come into play. Traditionally, ample attention has been given to natural climate variability, for which empirical data are available and probabilistic approaches can be used to estimate the likelihood and consequences of various climate scenario's, and to assess the effectiveness of alternative policies. It has been argued that under the rapidly changing global climate, the assumption of stationarity is no longer valid, meaning that historic return frequencies of extreme hydrological and weather conditions can no longer be relied on in water management: here the past is no longer the key to the future (Milly et al. 2008). Long-term decision making, involving, for example, decisions regarding water supply infrastructures having long lifetimes, brings a new class of important uncertainties. Climate change, itself being uncertain, may have significant – yet uncertain – impacts on precipitation and evaporation patterns, salt intrusion, salinization, and seasonal variation. In addition, a variety of uncertain but possible developments in land use and agricultural conditions, food markets, spatial developments, economic developments, future technology, and future societal values may also affect the future impacts and valuation of policies that have to be decided in the short term.

Partly but not exclusively triggered by the need to explore the impacts of climate change, during the last decade, the literature on dealing with various sorts of uncertainties has been developing rapidly (e.g., Maxim and Van der Sluijs 2011; Van Bree and Van der Sluijs 2014). Therefore, in this paper, we will first give a brief overview of some of the major developments in the uncertainty literature. Next, three examples are presented of application of one of the many approaches under development in the field of fresh water supply. We conclude with a few general observations on the state of the art, and recommendations for further work in the field.

2. DEVELOPMENTS IN DEALING WITH UNCERTAINTIES

Knowledge available for long-term policy making is unavoidably incomplete and often controversial. In the science-policy interface, strategic use of uncertainties seems to dominate: uncertainties are either downplayed to promote political decisions or they are overemphasized to prevent political action. Both are problematic and lead to policy strategies that are not fit for meeting the challenges posed by the uncertainties and complexities faced. While the scientific community has not ignored the problems of uncertainty, most efforts thus far have been associated with analysing and dealing with those uncertainties that can be described quantitatively (e.g., probabilistically, like in risk analysis), and with efforts at reducing uncertainties through additional research and/or refined modelling.

Not all uncertainties encountered in decision making can be quantified. In fact, unquantifiable uncertainties can sometimes be more relevant than the part where we do have enough knowledge to quantify uncertainty. Rather than using such unquantifiable uncertainties in a strategic way, or ignoring them, there is a growing consensus among scholars that deliberate policy strategies in the face of often irreducible uncertainties should be developed and implemented (Van der Sluijs et al 2008), and a wide variety of approaches and methods have emerged in the literature on the subject. As a result, the field is quite fragmented, and different key authors and groups have developed their own terminologies. A certain degree of conceptual confusion results, as approaches partially overlap, and different authors use the same terms in different ways. For example, approaches like 'Adaptive Management', 'Adaptation Policy Framework', and 'Robust Decision making' strongly (but not completely) overlap in their guiding principles.

Despite the fragmented character of the field, there seems to be a general agreement that different approaches are suitable under different degrees and types of uncertainty, that in real world situations often different types of uncertainty are present at the same time, and that combinations of approaches need to be fine-tuned to the characteristics of the situation. We will now briefly elaborate on these aspects.

2.1 Types and Levels of Uncertainty

In an attempt to harmonize diverse literatures in as far as they relate to model-based decision support, Walker et al. (2003) presented an uncertainty framework. The starting point for this framework is the distinction between an analyst's perspective on uncertainty and a decision maker's perspective on uncertainty (van Asselt 2000). The framework focuses on the analyst's perspective on uncertainty. A central idea of this framework is that uncertainty is a multi-dimensional concept. Ironically, this framework in turn had its own variants, motivated by various perceived shortcomings of the original typology or perceived need to tailor it to a specific field of study. More recently, Kwakkel et al. (2011) presented a review of the variants and what motivated them, and presented a new framework that integrated and harmonized the variants. Both the original and the new framework categorize uncertainties by their location, nature, and level. The location dimension focuses on where the uncertainty is located. The nature dimension focuses on the character of the uncertainty, i.e., whether the uncertainty is knowledge-based, a direct consequence of inherent variability, or related to human interpretation (ambiguity) or human values now and in the future. The level dimension focuses on the severity of the uncertainty.

Each of the three dimensions is relevant when selecting an appropriate approach for handling the uncertainty. However, the level dimension plays the most important role. Broadly speaking, the level of uncertainty is the assignment of likelihood to things or events. In some cases the likelihood or plausibility of these things, events or situations can be expressed using numbers, but in other cases more imprecise labels are used, such as more likely, less likely, or equally likely. Overall, the level of uncertainty ranges from complete certainty to absolute ignorance. We define five intermediate levels of uncertainty: we speak of level 1 uncertainty, or recognized uncertainty, when one admits that one is not absolutely certain but when one is not willing to measure the degree of uncertainty in any explicit way (Hillier and Lieberman 2001). We speak of Level 2 uncertainty, or shallow uncertainty, when one is able to enumerate multiple possibilities and is able to provide probabilities. We speak of Level 3 uncertainty, or medium uncertainty, when one is able to enumerate multiple possibilities and able to rank order them in terms of perceived likelihood. However, how much more likely or unlikely one possibility is compared to another cannot be specified. We speak of Level 4 uncertainty, or deep uncertainty, when one is able to enumerate multiple possibilities without being able or willing to rank order the possibilities in terms of how likely or plausible they are judged to be. Finally, we speak of Level 5 uncertainty, or recognized ignorance when one is unable to enumerate multiple possibilities, while admitting the possibility of being surprised (Kwakkel et al. 2011). Table 1 summarizes the five levels.

Table 1 The five levels of uncertainty (adapted from Kwakkel et al. 2010b)

Level of Uncertainty	Description	Examples
Level 1 (recognized uncertainty)	Recognizing that one is not absolutely certain, without being able or willing to measure the uncertainty explicitly.	Performing a sensitivity analysis on a parameter in a model by changing its default value with some small fraction.
Level 2 (shallow uncertainty)	Being able to enumerate multiple alternatives and being able to provide probabilities (subjective or objective)	Being able to enumerate multiple possible futures or alternative model structures, and specify their probability of occurring
Level 3 (medium uncertainty)	Being able to enumerate multiple possibilities and being able to rank order the possibilities in terms of perceived likelihood. However, how much more likely or unlikely one alternative is compared to another cannot be specified	Being able to enumerate multiple possible futures or alternative model structures, and being able to judge them in terms of perceived likelihood
Level 4 (deep uncertainty)	Being able to enumerate multiple possibilities without being able to rank order the possibilities in terms of how likely or plausible they are judged to be	Being able to enumerate multiple possible futures or specify multiple alternative model structures, without being able to specify their likelihood
Level 5 (recognized ignorance)	Being unable to enumerate multiple possibilities, while admitting the possibility of being surprised	Keeping open the possibility of being wrong or being surprised

2.2 A New Paradigm for Handling Uncertainty

Uncertainties pose a significant challenge to planning and decision making. The dominant approach in many fields has been to ignore the uncertainties, to quantify them into error margins, to try and reduce them, or to deal with only those uncertainties that can be easily quantified (Quade 1982; Dempsey et al. 1997; Marchau et al. 2009; Van Geenhuizen et al. 2007; van Geenhuizen and Thissen 2007; McDaniel and Driebe 2005). However, such approaches suffer from the problem that they focus on those uncertainties that are "among the least of our worries; their effects are swamped by uncertainties about the state of the world and human factors for which we know absolutely nothing about probability distributions and little more about the possible outcomes" (Quade 1982). Similarly, Goodwin and Wright (2010) (p. 355) demonstrate that "all the extant forecasting methods – including the use of expert judgment, statistical forecasting, Delphi and prediction markets – contain fundamental weaknesses." And Popper et al. (2009) state that the traditional methods "all founder on the same shoals: an inability to grapple with the long-term's multiplicity of plausible futures." In response to this, various new planning approaches have been put forward (e.g., Lempert et al.2003; Walker et al. 2001; de Neufville 2000, 2003; Dewar 2002; Dewar et al. 1993; Holling 1973; Lempert2002). These approaches generally contain three types of elements:

- analytic methods and tools to capture and assess the (consequences of) uncertainties,
- specific action or policy strategies in light of irreducible uncertainties,
- specific, often participatory processes to involve stakeholders and decisionmakers

Analytic approaches (Swanson et al. 2010) generally emphasize the need for a more thorough integrated forward-looking analysis of the uncertainties through techniques such as exploratory modelling and analysis (Agusdinata 2008; Lempert et al. 2003), bounce casting (Kahan et al. 2004), and scenarios in various forms (Bradfield et al. 2005; Varum and Melo 2010).

The action or policy strategies can be divided in three partly overlapping categories: First, so-called top-down or robust strategies build on analytic approaches in order to assess the vulnerability of alternative strategies, and to select those policies that do well under a wide variety of possible futures. Second, so-called bottom-up, resilience-based strategies recognize the limited capability of analytic approaches for anticipating rare events, and focus on the system-to-be-governed or to-be-designed itself, and try to enhance the system's inherent ability to cope with disturbances and its self-organizational capacity in a comprehensive way (e.g., Wardekker et al. 2010). Third, there is a growing interest in flexibility and adaptability in plans in which a strategic vision of the future is combined with short-term actions and a framework that can guide future actions (Albrechts 2004; Walker et al. 2001; Walker et al. 2013).

Together with a focus on involving stakeholders and decisionmakers in both the identification of uncertainties and the assessment of alternative strategies, these new approaches can be considered to form a new emerging paradigm for handling uncertainty differently. They are based on accepting the uncertainties, and focusing on the question what can best be done now, given that we don't know what the future will bring.

For a more complete overview of concepts, approaches and methods, the reader is referred to recent review reports and articles (Dessai and Van der Sluijs 2007, 2011; Wardekker 2011; Walker et al 2013; Lourenço et al 2014a; van Bree and van der Sluijs 2014).

2.3 Applications in the Field of Fresh Water Supply

As explained above, fresh water supply planning faces a variety of uncertainties, part of which cannot adequately be dealt with in a (purely) probabilistic way. Recently, a number of example applications of the type of approaches mentioned above have been reported in the literature. For example, Groves, et al. (2007;2012) have reported on a number of studies on fresh water supply in California and Colorado River, using the so-called 'robust decisionmaking' approach that is strongly based on analytical and computational techniques. Matrosov et al (2013a; 2013b) apply among others Robust Decision Making to a fresh water supply case in the United Kingdom. Groot et al. (2014) report, among other things, on several practical cases of water supply planning in Portugal, the UK, and Hungary. These studies mostly were related to dealing with uncertainties of the scenario-type, and used a combination of multiple methods such as expert elicitation, stakeholder involvement and model-based sensitivity analysis. The studies demonstrate, on the one hand, the complexity of the challenge, and, on the other, the relevance of more explicitly dealing with uncertainties, and 'a shift towards a flexible, robust and no-regret approach' (Groot et al. 2014, p 67). Lourenço et al. (2014b) also emphasize a preference for 'options that contribute to enhance resilience and adaptive capacity'.

In another study in Hungary (Malatinszky et al. 2013), the focus was on climate adaptation strategies for wetlands and grasslands. The authors emphasize the importance and various benefits of stakeholder involvement in such processes, which include "enhanced awareness, willingness to taking action, inclusion of local knowledge, information exchange among affected parties, identification of win-win-solutions for land users and nature conservation, and building trust in authorities".

Clearly, while there seems to be agreement on the general tendencies of preferred approaches, much remains to be developed and learned, both with respect to methods and tools and their potential and limitations in practical applications and with respect to what strategies fit what uncertainty situations.

In the following sections, we will therefore, in order to learn more about their applicability, illustrate the application of three novel but different approaches to two different real-world fresh water supply cases in The Netherlands. First, the resilience approach is applied to an area of peat grassland in the west of the Netherlands. Second, a system robustness approach is applied to a low-lying polder area mainly used for agricultural purposes. Third, an exploratory modeling approach is applied to the same case study area.

3. A RESILIENCE APPROACH: THE VENEN-VECHTSTREEK REGION

3.1 Resilience

Resilience is a concept that emerged in ecology research in the 1960s. It was used in relation to the stability of ecosystems in the face of a wide range of external perturbations: "the amount of disturbance that a system can absorb and still remain within the same state or domain of attraction" (Holling 1973; Walker and Salt 2006). The concept has since been adopted by various disciplines, ranging from engineering to psychology, disaster studies, and climate change adaptation. Walker et al (2004) define resilience in relation to 'social-ecological systems' as: "the capacity of a system to absorb disturbance and reorganize while undergoing change so as to still retain essentially the same function, structure, identity, and feedbacks". Resilience focusses on coping with disturbances and preventing collapse into a qualitatively different state, rather than predicting and preventing disturbances.

In relation to climate change adaptation, resilience can be understood as a bottom-up approach, which aims at strengthening the system, removing vulnerabilities, and enhancing the capacity to cope with and recover from climate change-related disturbances and surprises. The case-study examined the consequences of various proposed measures and developments on the climate-resilience of the freshwater systems in the Venen-Vechtstreek region. These measures are plans to manage the area in general and are not specifically proposed to adapt to climate change.

Resilience analysis can take several steps. The first step is to determine 'resilience of what?'. This includes an overview of the area, the trends and challenges it faces, the way it is used, and the key characteristics and functions that make the system for what it is (i.e., if these are not retained, the system has shifted to a different state). The second step is to address the question of 'resilience to what?'. In this study, an inventory of climate change impacts was made. The third step is to assess the resilience of the system, measures that will enhance resilience, or the resilience consequences of developments. This case study concerned the latter. We applied a set of 'resilience principles' describing different mechanisms by which systems can absorb disturbances and retain identity (Watt and Craig 1986; Barnett 2001; Wardekker et al.2010):

- Homeostasis: multiple feedback loops counteract disturbances and stabilize the system.

- Omnivory: vulnerability is reduced by diversification of resources and means.

- High flux: a fast rate of movement of resources (e.g., information, money, expertise, goods for emergency relief, fuel) through the system ensures fast mobilization of these resources to enable coping with perturbations through for instance reduced feedback delay. If information is proactively and rapidly exchanged, parties can act on this faster when there is a problem.

- Flatness: the hierarchical levels relative to the base should not be top-heavy. Overly hierarchical systems are too inflexible and too slow to cope with surprise. The more hierarchical levels exists through which stabilizing feedback signals must operate, the lower the signal to noise ratio and the lower the speed of signal transmission, and thus of action to counteract the perturbation or remedy the damage.

- Buffering: essential capacities are over-dimensioned such that critical thresholds are less likely to be crossed.

- Redundancy: overlapping functions; if one fails, others can take over.

3.2 Case Study Area and Trends

The Venen-Vechtstreek region is an area of peat grassland in the west of the Netherlands. It is known for its wide views, cow-filled meadows, and long and narrow stretches of land separated by water (Fig. 1). The appearance of the region has changed little over the centuries. The case study focused on the specific area of Groot Wilnis – Vinkeveen. The region is part of a range of wetlands forming a 'robust ecological corridor' between areas protected under Natura 2000, the European ecological network of protected areas. The corridor is intended to provide a connection and shelter that allows animals and plants to migrate between nature areas. The area is situated at approximately 2.5 m below sea level and is artificially drained.

Due to the artificial drainage, peat comes into contact with oxygen and decomposes. Consequently, the area suffers from soil subsidence. The most sensitive parts of the area have seen subsidence of up to 12 mm per year. Agricultural areas suffer more subsidence than nature areas, due to higher levels of drainage, resulting in large local differences in water levels. The decomposing peat also releases nutrients into the surface water and greenhouse gasses into the atmosphere.

The area is valuable for both ecological and agricultural reasons. The area's management covenant (Stichting Ontwikkeling De Venen 2010) specifies that the area should be preserved as an open landscape in which the dairy sector can continue to develop. Four key functions can be distinguished: fresh water supply, nature, agriculture, and recreation. Stakeholders aim to strengthen these functions, as well as to reduce the amount of soil subsidence. The functions have different implications for the management of the area. Connecting them through integrated measures will be an important step in enhancing resilience.

Figure 1. Satellite view of the case-study area (source: Google Maps)

Climate change will affect Dutch freshwater systems through increasing air and water temperatures, sea level rise, changes in seasonal river discharge, and changes in precipitation (PBL 2013). Average precipitation may either increase or decrease in summer, and increase in winter. For the study area, this may result in an increasing potential precipitation deficit during summer. Dry periods could negatively impact water availability and quality for fresh water supply as well as for nature and agriculture. Wet periods could result in flooding, negatively impacting all four functions. They could also exacerbate the current tension in water level requirements for agriculture (relatively low) and nature (relatively high), and enhance nutrient leaching from fields into surface water, negatively impacting nature. Temperature changes can impact various plants and animals. For

2050, a scenario study projects a moderate to high terrestrial drought risk, moderate risk of aquatic eutrophication, low risk of terrestrial eutrophication, and low to high risk of aquatic salinization (Verhoeven 2012). The high salinization risk occurred in a scenario with warm, dry summers, when brackish water from outside the area would be required to manage the water levels.

3.3 Analysis

A list was drafted including measures that are currently planned for managing the study area, based on the management covenant, other policy documents, and position papers of regional and national stakeholders. This list was supplemented with options that could contribute to the area's overall resilience. These supplementary options were collected through interviews with stakeholders and experts with experience concerning the study area. Options were categorized according to the key functions they contribute to: nature, agriculture, water quality, and recreation.

The research team performed an initial reflection on the resilience implications of the options, followed by a second set of interviews. These reflections were used in preparation of a workshop with stakeholders and experts on the area and on resilience: to select a subset of options to be discussed in the limited time available, and to seed the discussion. Workshop participants ($n = 7$; this number is adequate for an expert assessment; 6–12 is advisable, to cover relevant fields of expertise, but allowing for sufficient time for discussion; Knol et al. 2010) scored the shortlist of options on each of the resilience principles. They applied a five-point scale, ranging from strong decrease to strong increase of resilience. Aggregated 'resilience scores' were also generated. See Fig. 2.

Nearly all of the assessed options for nature improved resilience, according to the scores of most participants (occasionally, one or two participants disagreed). The participants expected the option 'marshland construction and capillarity' to perform particularly well. Agricultural options scored less well; half resulted in (slightly) reduced resilience. The option 'underwater drainage' did receive a good score. For both types of options, the range between individual aggregated scores was generally one to two points. Scores on separate resilience scores had similar ranges. They could on occasion diverge more strongly (up to scores of -2 to +2, cf. Fig. 2, although these were exceptions), but the interquartile ranges (25–75 percentiles of scores) were two points in most cases. The group size was sufficiently large that a single expert scoring notably different did not change the interquartile ranges.

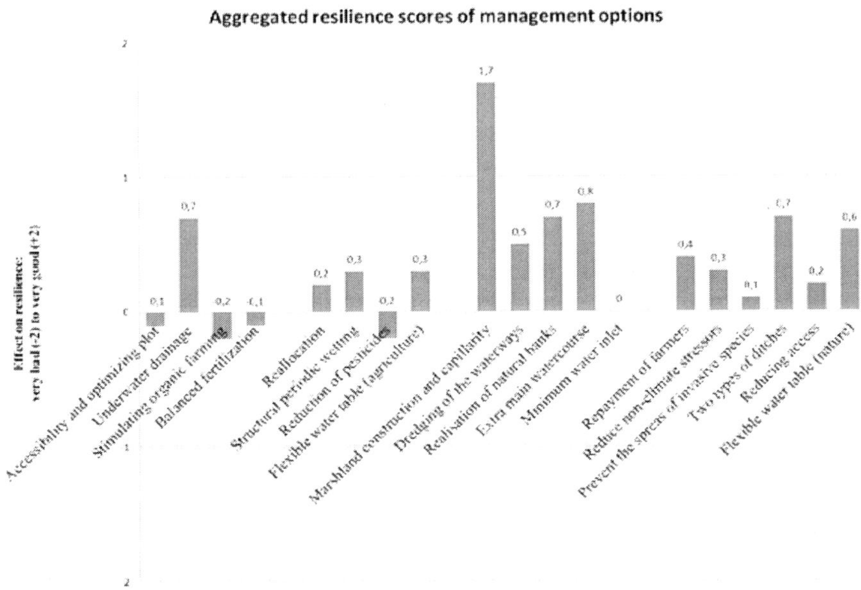

Figure 2. Mean aggregated resilience scores. Options are arranged in four groups, from left to right: agriculture (planned and new) and nature (planned and new)

The approach to assessing resilience applied in this case study provided a useful tool to scan a variety of options on a range of impacts for a range of stakeholders on their implications. It allows for a relatively quick evaluation of options and the inclusion of stakeholders and their perspectives. The approach was also easy to combine with other decision criteria in a multicriteria assessment, allowing for a broader evaluation of policy options. Preferably, this would include multiple workshops with a broader range of expertise amongst the participants. A downside is that the assessment takes place on an ordinal scale (categories, rather than full quantification) and does not examine spatial patterns. Also differences in opinion between participants scores will remain, even with extensive discussion. These require further investigation and may hint at deeper differences in opinion (uncertainty due to ambiguity; see e.g., Brugnach et al. 2008), or different ways in which an option can be implemented. The approach can be applied for a broad range of cases, and does not require extensive computing power or special software. It does require some insight in the system dynamics of the case study (among researchers and participants), as well as experience with participatory processes (among researchers). It is possible to scan a broader set of options with the research team only, and/or include a larger set of stakeholders by means of a survey. However, discussion with, and among, local stakeholders and experts remains important to analyze local resilience in a meaningful way.

We concluded that, considering that agriculture is a key function of the area, covering most of the land use, it would be advisable to pay attention to increasing resilience in agricultural developments. Scores on separate resilience principles (i.e., not aggregated into a single resilience score) showed that very few options increase flatness. Additionally, few agricultural options address redundancy.

4. A SYSTEM ROBUSTNESS ANALYSIS: CASE RIJNLAND

4.1 System Robustness

System robustness is a concept that originates from engineering and biology, where it refers to the ability of systems to maintain system characteristics when subjected to disturbances (Carlson and Doyle 2002; Stelling et al. 2004). For example, a road network is robust to incidents when the effect on traffic congestion is prevented and/or limited (Snelder et al. 2012). According to the Merriam-Webster dictionary, the term*robustness* indicates that systems or networks are capable of performing without failure under a wide range of conditions. In the context of flood and drought risk management, system robustness is defined as the ability of a socio-economic system to remain functioning under a range of disturbance magnitudes (Mens et al. 2011). The term disturbance refers to an external event resulting from climate variability, for example floods or droughts. The system for which robustness is assessed comprises biophysical as well as socio-economic aspects. For example, the robustness of a flood risk system does not only depend on the flood protection system, but also on the characteristics that determine the flood impact, and on the social and economic capacity to recover from these impacts. Failure of the system thus means that impacts are too large to recover from. Such impacts may be considered a catastrophe or a disaster. Likewise, a robust drought risk system is one that can resist frequent droughts (zero impact) and that can recover from socio-economic impacts of rare droughts (limited impact).

System robustness can be assessed by studying the so-called response curve: the relationship between disturbance and response, for example the relationship between precipitation deficit and crop yield deficit. Key aspects of the response curve for this application are:

- Resistance threshold: the smallest deficit causing yield deficit;
- Manageability: the severity of the yield deficit in comparison with a recovery threshold (i.e., the societally unacceptable level of yield deficit).

Manageability refers to how severe the impacts of a drought can become. Whether a response is considered severe depends on the social and economic capacity to recover from the response to a disturbance (see De Bruijn 2004). This capacity can be quantified by a recovery threshold, indicating the level of response from which it will be very difficult to recover. For example, when economic impacts of flooding exceed 5 % of the national GDP, it often means that aid is needed from other countries (5). Manageability indicates the extent to which the response stays far from the recovery threshold for the range of considered disturbance magnitudes.

4.2 Case Rijnland

Droughts occur as a natural temporary feature of the climate (AMS 2013), but it may have an effect on reservoir storage, groundwater resources and river discharges, thereby potentially causing socio-economic and environmental impacts. Climate change may cause a change in the frequency and severity of droughts. Many water resources studies focus on the impact of climate change on long-term average flows (Lehner et al. 2006), thereby analysing the average demand and supply balance in the long-term. However, even if water supply is sufficient to meet demand on average, droughts may cause a temporary shortage of water which may have far-reaching impacts on society. Insight is thus needed into the impact of potential drought events resulting from climate variability, now and in the future.

To understand how systems robustness analysis can aid in the management of drought risk systems, it was applied on a case: Rijnland polder areas (Mens and Klijn 2015; Mens et al. this issue). A drought risk system is a geographical area that depends on one or more water resource for several functions (e.g., drinking water supply, industry, nature and agriculture). Drought, in this case defined as a period of precipitation deficit (i.e., evapotranspiration is larger than precipitation), may cause water shortage with an impact on the area's functions. Precipitation deficit is thus considered the disturbance to the system.

The case study area, Rijnland, is a system of low-lying polders with agriculture (potatoes, horticulture, bulbs, flowers, dairy farming) as the main land use type. This is a mixture of rain-fed and irrigated agriculture. The water supply and drainage system consists of ditches, canals, lakes, and pumps and sluices. In winter, there is an excess of water that is drained and pumped out, and in summer fresh water is let into the surface water system from a nearby river (see also Mens et al. this issue). The inlet water is used for water level control, flushing for water quality management and water supply (irrigation, drinking water, industry). Water level control guarantees the stability of canal embankments, and flushing reduces the salinity in the canals and ditches, caused by seepage of saline groundwater. During an average summer, the total external fresh water inlet amounts 40 to 60 10^6 m^3. In a dry summer, this may increase to about 100 10^6 m^3 (Rijnland2009). The analysis focuses on the impact of drought on agriculture.

4.3 Model

Figure 3 presents an overview of the model structure. The water demand module generates water demands for irrigation and water level control in rural areas and is a simple two layer grid-based groundwater model with a resolution of 250 by 250 m, taking into account a limited number of land use and soil types. For each layer in each grid cell, the model calculates the water balance. First, the potential evaporation is calculated by multiplying the reference evaporation with a crop factor that is specified for each crop and 10-day period. The actual evaporation is a function of the potential evaporation, the moisture in the root zone, and the soil moisture suction (pF value). Lateral flow from groundwater to local surface water and vice versa is a function of groundwater depth relative to surface water level. Water flowing from the root zone to the subsoil (percolation) depends on the root depth, porosity, and precipitation. Capillary rise (flow from subsoil to root zone) is calculated as a function of the groundwater depth below the surface level and the root zone suction (Kabat and Beekma 1994; Oosterbaan 2001). The lower boundary condition of each plot is an annual seepage flux taken from results of the complex model for an average year. In case the root zone and subsoil are saturated, excess water is moved through surface runoff. In urban areas surface runoff is a function of the net precipitation and a runoff co- efficient of 0.8 (Urbonas and Roesner 1993). The water demand is determined from the difference between the actual and potential evaporation. The amount of water requested for maintaining the target water level in the local surface waters areas is derived from the net precipitation and the surface area of these waters. The grid cells are aggregated over a watershed area (called district).

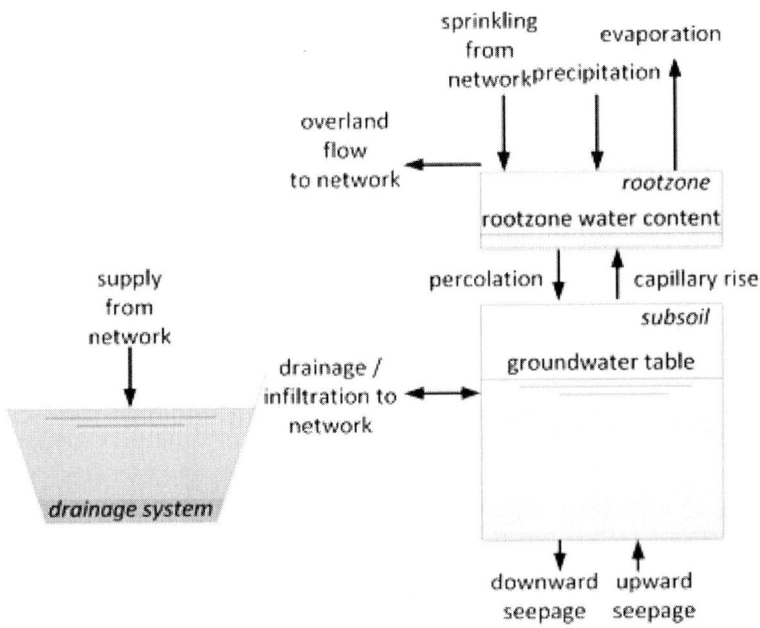

Figure 3. Diagram showing the basic structure of the model (adapted from Haasnoot et al. (2014))

The salt intrusion module simulates the salt concentration at the Gouda inlet depending on river discharge and sea level. This module is based on empirical correlation between the Rhine discharge at Lobith and salt concentrations in the lower river reaches calculated using a 1D hydraulic model (SOBEK) (van den Boogaard and van Velzen 2012).

$$Salt = 1700 + (90-1700) \times \frac{e^{Fact}}{1+e^{Fact}}$$
$$Fact = \left(\frac{Q_{lobith}-600}{2.211}\right)^{0.309}$$

where Q_{lobith} is the discharge at Lobith in cubic meter per second and *Salt* is the salt concentration at the Gouda inlet in milligram per liter. As discussed in Haasnoot et al. (2014), this relation will slightly underestimate the frequency of closure at Gouda.

The focus of this analysis is on the economic damages to agriculture due to drought. For this, we use Agricom (Mulder and Veldhuizen 2014) which is an agro-economic model to estimate agricultural yield losses due to water shortage, saline soil moisture and water excess. Drought is defined in terms of

$$E_{ratio} = \frac{ET_{act}}{ET_{pot}}$$

Where ET_{act} is the actual evatranspiration and ET_{pot} is the potential evapotranspiration. Given E_{ratio} and crop specific damage curves that take into account the growing season, the E_{ratio} is translated into a loss of yield in kilogram, which in turn is monetized. In our analysis, we calculate both the potential yield in Euro, assuming perfect conditions, and the actual yield in Euro's, allowing us to calculate the relative loss due to deficiencies in the system.

4.4 Results

The response curve of the reference situation is given in Fig. 4. Each point in the curve is a sample from climate variability according the current climate and according to the worst-case climate change scenario for 2100. Together they provide an indication of the range of drought magnitudes that may occur and the corresponding impact on agricultural yield deficit.

The vertical range is explained by the timing of the precipitation deficit within the growing season. Some crops are more sensitive to drought in the beginning of the season, while others are more sensitive at the end of the season. Droughts with similar precipitation deficits may therefore occur, but when they differ in timing the impact will differ as well.

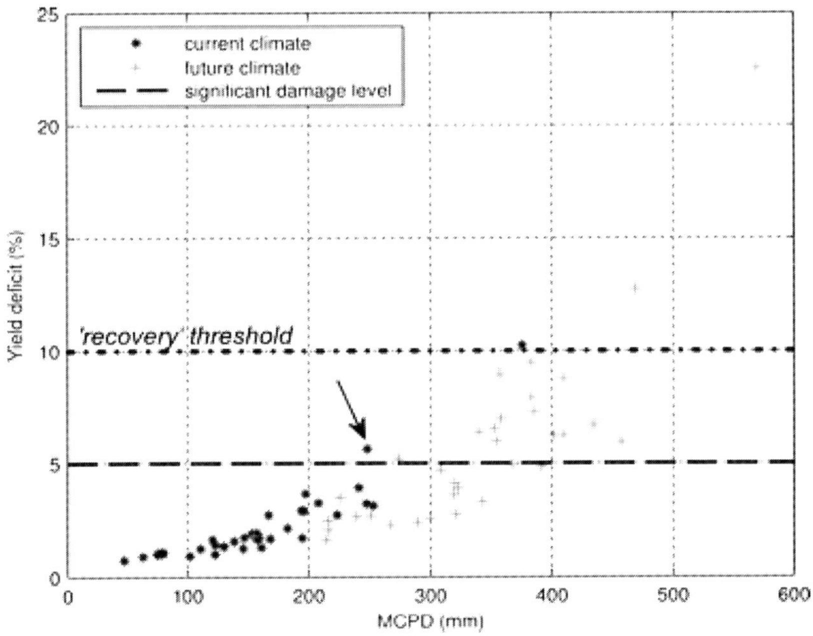

Figure 4. Response curve of the Rijnland system: yield deficit as a function of maximum cumulative precipitation deficit (MCPD), the indicator for drought magnitude

There are two lines drawn in the figure: one at 5 % yield deficit and one at 10 % yield deficit. Yield deficits below 5 % are considered not significant, because farmers take into account some level of business risk. The first point that exceeds this line (indicated with an arrow) indicates the *resistance threshold* at 250 mm precipitation deficit. The 10 % line is arbitrarily chosen to indicate the level of unacceptable yield deficit that must be avoided (recovery threshold). Only two of the 35 events exceed this 10 % line in the future climate. When it is aimed to enhance the *manageability*, measures should be designed that reduce the impact for these events.

In practice, the location of both the 5 % and the 10 % line is a political choice. In fact, the real recovery threshold in this system will never be exceeded, since it can be expected that the area can still economically recover from 100 % yield deficit (see Mens et al. this issue). The Rijnland system can be considered robust to droughts, because the maximum simulated yield deficit does not exceed 25 % for a large range of precipitation deficits. This can be partly explained by the storage capacity as well as the diversity of crop types that vary in when they are most sensitive to water shortage (Mens et al. this issue).

4.5 Reflection

Although the model obviously is a limited representation of reality, it suffices for the purpose of the robustness analysis. Different model parameter assumptions could slightly change the resistance threshold, but the score on proportionality and manageability is not expected to change. Thus, the conclusion on how well the system is able to deal with a range of drought events is not affected by the choice of the model.

Two types of measures are often considered in a drought risk context: those that increase the water supply and those that reduce the water demand. In traditional decision making approaches, where supply reliability is the main decision criterion, both types of measures will increase the supply reliability. However, a high reliability does not inform about the consequences of low-probability drought events. Additional criteria are thus needed that assess measures on how they affect the system in terms of avoiding large societal and economic impacts due to droughts. The two robustness criteria, resistance threshold and manageability, seem to fulfil this need. More applications are needed to explore how different types of drought risk reduction measures score on the robustness criteria.

We believe that a robust system, that can deal well with frequent as well as rare events, will also be able to deal well with changed variability under a future climate (Watts et al. 2012). Knowledge or assumptions about the future climate variability are not needed to perform robustness analysis. However, climate change scenarios are helpful in determining the range of drought events for which the system's robustness is assessed.

5. AN EXPLORATORY MODELING APPROACH FOR DESIGNING ADAPTATION PATHWAYS

5.1 Adaptation Pathways

An emerging approach to the design of climate adaptation plans and strategies is adaptation pathways (Haasnoot et al. 2013; Wise et al. 2014). This approach combines two bodies of literature on planning under uncertainty: work on adaptive policymaking (Walker et al. 2001; Kwakkel et al. 2010; Hamarat et al. 2013) and work on adaptation tipping points and policy pathways (Kwadijk et al. 2010; Haasnoot et al. 2012; Offermans 2012). A plan is conceptualized as a series of actions taken over time. In order to come to a good plan, it is necessary to first identify candidate policy actions and their adaptation tipping points. An adaptation tipping point is the condition under which a given policy action no longer meets its objectives. The point in time at which this happens (the so-called sell-by-date) is scenario dependent. In light of an analysis of the sell-by dates of various policy actions, concatenations of options, or policy pathways, can be specified where a new policy option is activated once its predecessor is no longer able to meet the definition of success. Typically, there is a portfolio of pathways that decision-makers would like to keep open for the future. This adaptation map forms the basis for the plan. Figure 5 shows an example of such a map. For a more detailed elaboration on DAPP, see Haasnoot et al. (2013).

Model-based decision support for the design of adaptation pathways is challenging for a variety of reasons. There exist a wide variety of candidate policies that has to be considered. These policies might be sequenced in different ways, taking into account co-dependencies between the various options. Their assessment needs to explicitly take into account the dynamics over time, rather than some endpoint. The assessment also should cover the variety of outcomes of interest the different stakeholders involved in the decision-making care about. And last but not least, the assessment should cover the variety of irreducible uncertainties that exist and the performance of candidate pathways needs to be assessed over the full range of uncertain factors.

One approach that might be amendable to supporting the design of adaptation pathways is exploratory modeling. Weaver et al. (2013) have argued that adaptive planning requires an exploratory modeling approach. In exploratory modeling, modelers account for the various unresolvable uncertain factors by conducting series of computational experiments that systematically explore the consequences of alternative realizations of the various uncertain factors (Bankes et al. 2013). These computational experiments can be used also a test bed for candidate policies (Lempert et al. 2003).

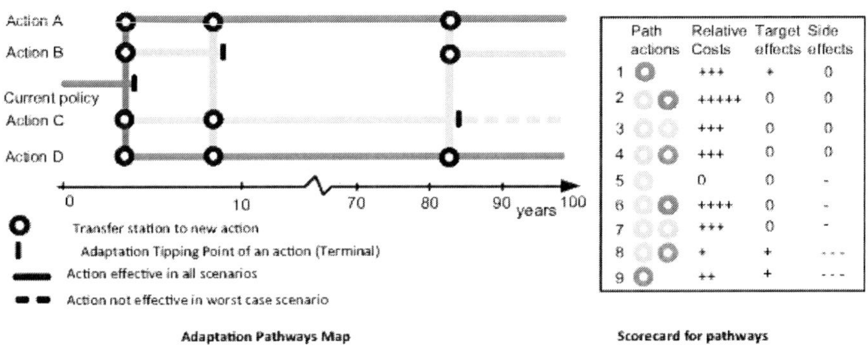

Figuer 5. An example of an Adaptation Pathways map (left) and a scorecard presenting the costs and benefits of the 9 possible pathways presented in the map. In the map, starting from the current situation, targets begin to be missed after 4 years. Following the grey lines of the current plan, one can see that there are four options. Actions A and D should be able to achieve the targets for the next 100 years in all climate scenarios. If Action B is chosen after the first 4 years, a tipping point is reached within about 5 years; a shift to one of the other three actions will then be needed to achieve the targets (follow the *orange lines*). If Action C is chosen after the first 4 years, a shift to Action A, B, or D will be needed after approximately 85 years in the worst case scenario (follow the *solid green lines*). In all other scenarios, the targets will be achieved for the next 100 years (the *dashed green line*). The colors in the scorecard refer to the actions: A (*red*), B (*orange*), C (*green*), and D (*blue*)

5.2 The Case, the Model and the Associated Uncertain Factors

As case, we use the Rijnland case as described in section 4. We use the same model as used there with a few small changes. First, we adapted the model to account for the following uncertain factors:

- River runoff in the Rhine
- Rainfall
- Land use
- Evatranspiration

For each of these sources of uncertainty we use transient scenarios (Haasnoot et al. 2014) to address them. So, rather than looking at the system in e.g., 2050, we consider the change over time from the current system to the state of the system in 2100. For river runoff, rainfall, and evapotranspiration we use different possible realizations of the climate scenarios W+ and G.. These are climate scnearios for the Netherlands and have been developed by the Dutch Royal metrological institute. W+ is an extreme warming scenario, while G represents a modest warming scenario (KNMI 2006). These realizations are the same as those used by Haasnoot et al. (2014), although here we only use the parts relevant to the case study area. We use 10 possible realizations of both scenarios. For land use, we use the land use as described in the four delta scenarios as developed in the Deltaprogram. These scenarios are labelled Warm, Pressure, Rest, and Steam. These delta scenarios combine the 2006 W+ and G climate scenarios (KNMI 2006) with socio-economic scenarios. Figure 6 shows the scenario logic for the scenarios. Maps were available for 2050 and 2100. We interpolated the maps in between. Given that we sum up over the region, a rather simple and crude interpolation has been used. Combining the different realizations with he delta scenarios gives 40 scenarios. To assess the role of changing land use, we include, in addition to the changing land use also a no change case, giving 60 scenarios in total (i.e., Warm, Pressure, Rest, Steam, No change W+, and no change G). Note that in the current results land use has some influence on water demand, but the irrigation maps do not evolve with land use.

5.3 Results

A variety of analyses have been carried out. As a first experiment, we explore how the system responds if water supply is deliberately closed. To this end we simulate two situations. In the first situation, we keep the water inlet at Gouda open at all time, in the other, we close the inlet at all times. We assessed the performance for both situations by analyzing the results for 5000 experiments. A boxplot for the loss of income of farmers due to reduced yields in both situations is shown in Fig. 7. So, this is the monetized loss of yield as explained near the end of section 4.3. As can be seen, there are a few more extreme cases in terms of income loss when the Gouda inlet is closed at all times. Still, the differences are quite small and it is relevant to investigate this further.

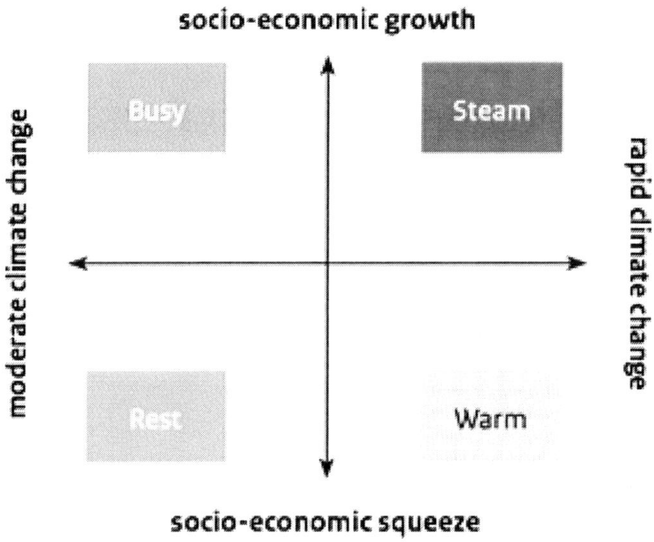

Figure 6. The scenario logic of the delta scenarios (adapted from (Delta Programme 2012)

In order to understand why there is only a minor difference in income loss between the situation where the fresh water inlet at Gouda is always open, and the situation where the fresh water inlet at Gouda is always closed, we looked at the worst-case delta scenario. In this scenario, warm, the system experiences severe climate change, combined with substantial land use developments putting even more pressure on the system. We explored the behavior of the system for all ten possible realizations. The resulting water demands per decade are shown in Fig. 8.

In Fig. 8 we compare the water demand at a decade level for only those situations where Gouda is closed on the one hand, and for all decades on the other. As can be seen, during the decades during which the fresh water inlet at Gouda is closed, the water demand is quite small. In fact, the demand is below the supply available via the small scale water supply (5 cubic meter per second). As such, no severe water shortages occur. This suggests that decades with high demand for fresh water do not coincide with low runoff off of the Rhine. In light of this analysis, it appears that no future adaptation actions are necessary. There are a few caveats with this conclusion.

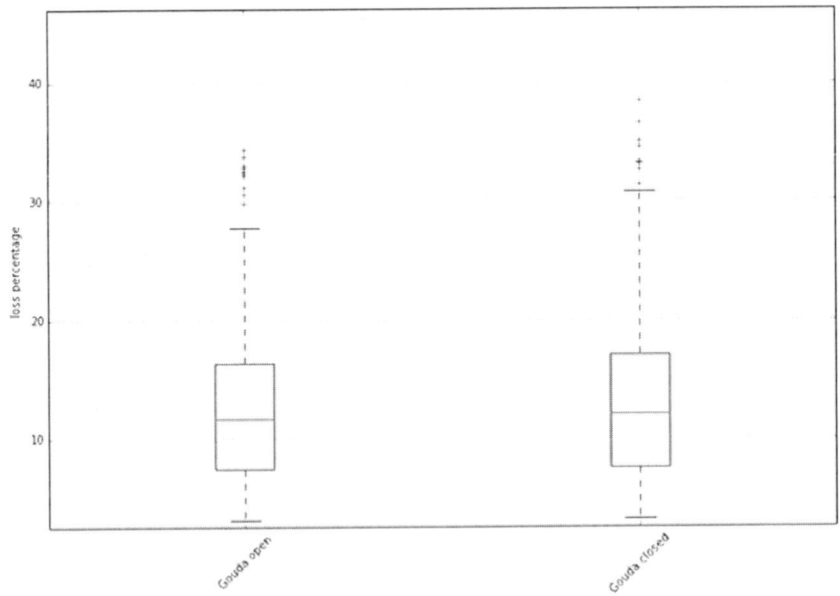

Figure 7. Boxplot of percentage income loss for farmers due to a reduction in yields, comparing the situation where the fresh water inlet at Gouda is open at all times with the situation where the fresh water inlet at Gouda is closed at all times

First, the relation used to calculate closure of Gouda is known to underestimate closures (Haasnoot et al.2014). As such, a closer analysis on a different time scale might be needed. The current analysis uses decades, and it might be necessary to shift to a daily analysis to get a better insight into what is happening.

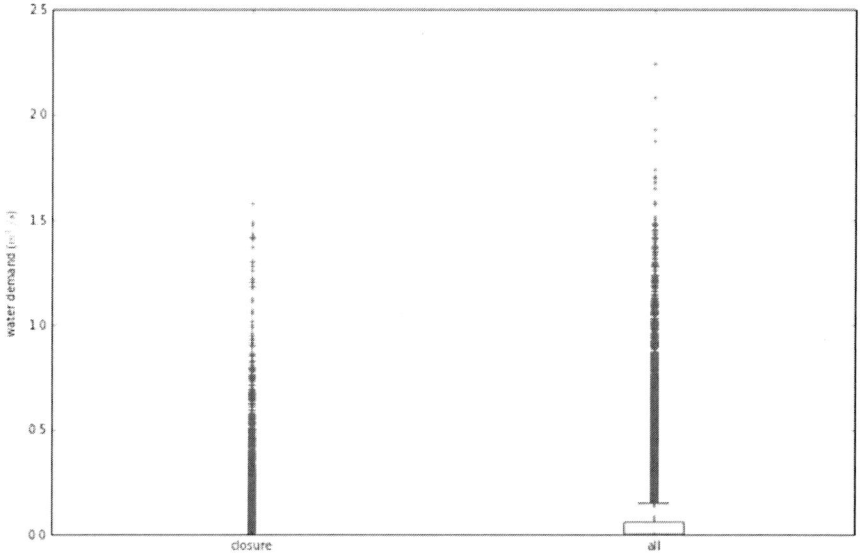

Figure 8. Water demand per decade for the Warm scenario. The left hand boxplot contains the water demand when the fresh water inlet at Gouda is closed because of salt intrusion. The right hand boxplot contains all the decades irrespective of whether the fresh water inlet at Gouda is closed or not

Second, the model does consider land use change. This land use change affects water demand. We have not translated the changes in land use into changes in the areas being irrigated. Doing so is possible. Ter Maat et al (2013), for example, introduce several additional assumptions regarding future irrigation demands, allowing the translation of land use change into changes in water demand. We speculate that one of the reasons for the differences between the results reported here and those reported in ter Maat et al. (2013) are due to how changes in land use affect irrigation.

Third, it appears that the model at every time step resets the water levels to the norm height. Inspection of the source code of the model did not provide clarity as to where the water for this comes from. That is to say, it appears that a hidden assumption in the model is that that the full capacity of the KWA is available for maintaining water levels. Based on ter Maat et al. (2013), this assumption is questionable. If the model were modified such that the KWA is first used to maintain water levels, and only subsequently if possible for irrigation, it is plausible that there would be more severe fresh water shortages for irrigation. In turn, this would increase the income loss of farmers.

From a methodological point of view, the main conclusions of the analyses reported on are that it is possible to consider a wide range of different uncertainties simultaneously. Exploratory modeling offers an approach for systematically mapping out the consequences of various uncertainties. We also demonstrated that it is possible to do a variety of potentially useful further analyses on the exploratory modeling results. On the one hand, these analyses provide usefull information for policy-making. On the other hand, the analyses create new insight into the behavior of the models. Moreover, these kinds of analyses can also help in assessing the usefulness of models, since they stress test the model and clarify the domain over which the model provides meaningful results.

6. CONCLUSIONS AND DISCUSSION

Long term planning for fresh water supply faces a variety of sources and types of uncertainty, many of which cannot be adequately dealt with using probabilistic approaches. In response, a variety of complementary approaches has been proposed and developed to explicitly recognize and deal with these uncertainties. On the one hand, analytical methods are being developed to capture and assess uncertainties and their impacts, on the other, various action strategies are proposed in view of uncertainties, including approaches based on robustness, on resilience, and on flexibility and adaptation. We have illustrated the application of respectively a resilience approach, a robustness approach, and an exploratory modeling approach. Reflecting on these different approaches, we observe that the robustness approach requires quantification of uncertainties and computational analysis. The exploratory modeling approach builds on computational power and enables exploration of a broad set of possible assumptions about the future while relaxing requirements of probabilistic specification.

The resilience approach, on the other hand, is primarily qualitative. As a result, the effectiveness and efficiency of resilience options is difficult to assess in quantitative terms. But the resilience approach may, as illustrated, be more suitable for tailoring adaptation to local situations, and appeal to notions of stakeholder involvement and ownership, self-governance and self-relience. In contrast to the robustness and exploratory modelling approach, the resilience approach can be used for those adaptation policy challenges where the state of scientific knowledge does not (yet) allow a sufficiently adequate form of quantification of impacts at the spatial and temporal scales that are pertinent to that particular adaptation policy challenge.

Application in the Venen-Vechtstreek case illustrates that the resilience approach enables involvement of multiple stakeholder perspectives and consideration of multiple societal reactions (that in turn can influence the effectiveness of the various management options) as well as a range of other criteria relevant to the decision making challenge such as costs and co-benefits of the management options. In our workshop, the co-benefits turned out to be crucial to obtain support for resilience-based climate adaptation measures from local stakeholders such as farmers. The approach is especially useful for a quick screening of management options.

The robustness approach focusses on a partial challenge around a single metric (vulnerability of the present fresh water system from the view point of the water manager only), but enables spatially explicit detail and quantitative assessment of societal costs and benefits often required in policy making contexts to justify interventions that affect various stakeholders in various ways.

In practical terms, the resilience approach is useful in the phase where long term management and adaptation strategies for a region are being designed in multiple stakeholder processes and considering multiple impacts on multiple sectors. Robustness analysis is more useful to assess specific issues (vulnerability to a specific impact for a specific sector), and fit for situations where the relevant set of future scenario's is known and quantifiable. As such the approaches can be highly complementary (see also Wardekker et al. 2010). Especially in cases where decisions on dimensioning of options is key, a combination of approaches can have added value: if a resilience based screening of options reveals the wish to create fresh water buffers in an area (buffering) or to increase the inlet of fresh water from outside the area (ominvory), then a specific robustness analysis can help to answer the questions how big the buffer should be or what is the required capacity of the extra fresh water inlet should be.

More generally speaking, in practical cases different types of uncertainties will play a role, and analyses and strategies at different levels will be needed. As a result, a tailored combination of different approaches may be the best way forward, combining more top-down measures with bottom-up approaches. Clearly, both the analytical approach and the choice of strategies will have to be tuned to the characteristics of the situation, and more research and experience are needed to increase our understanding of the advantages and disadvantages for various approaches, and how they relate to the various types of uncertainties.

Another lesson from the cases presented here is that the match between the choice of policy objectives (which possible impacts do we seek to reduce with the adaptation policies) and the choice of scope and system boundaries of the scientific assessment carried out to support that decision making process is crucial. Is the key concern in agricultural impacts the *physical* yield (in kg of crops per ha) or the *economic*yield (in monetary units per ha)? Should we focus on the robustness of water *supply* only (how can we maintain the fresh supply under future climates to meet the given future demand) or should we address the robustness of the *entire* fresh water system? In the latter case, adaptation of water *demand* to a reduced supply would also need to be considered in the scientific assessment. This relates to the question whose perspective should guide the scientific assessment and who is and is not deemed responsible for adaptation.

REFERENCES

1. Agusdinata DB (2008) Exploratory modeling and analysis: a promising method to deal with deep uncertainty. Ph.D. thesis, Delft University of Technology

2. Albrechts L (2004) Strategic (spatial) planning reexamined. Environ Plann B 31:743–758

3. AMS (2013) Drought. An information statement of the American meteorological society. https://www.ametsoc.org/POLICY/2013 drought_amsstatement.html. Accessed 29 September 2014

4. Bankes SC, Walker WE, Kwakkel JH (2013) Exploratory modeling and analysis. Encyclopedia of operations research and management science, 3rd edn. Springer, Berlin

5. Barnett J (2001) Adapting to climate change in Pacific Island Countries: the problem of uncertainty. World Dev 29:977–993

6. Bradfield R, Wright G, Burt G, Cairns G, Van Der Heijden K (2005) The origins and evolution of scenario techniques in long range business planning. Futures 37:795–812

7. Brugnach M, Dewulf A, Pahl-Wostl C, Taillieu T (2008) Toward a relational concept of uncertainty: about knowing too little, knowing too differently, and accepting not to know. Ecol Soc 13(2):30

8. Carlson JM, Doyle J (2002) Complexity and robustness. Proc Natl Acad Sci 99:2538–2545.

9. De Bruijn K (2004) Resilience indicators for flood risk management systems of lowland rivers. Int J River Basin Manag 2:199–210

10. De Neufville R (2000) Dynamic strategic planning for technology policy. Int J Technol Manag 19:225–245

11. De Neufville R (2003) Real options: dealing with uncertainty in systems planning and design. Integr Assess 4:26–34

12. Delta Programme (2012) Delta Programme 2013: the road towards the Delta Decisions, Dutch Ministry of Infrastructure and the Environment and the Ministry of Economic Affairs, Agriculture and Innovation

13. Dempsey PS, Goetz AR, Szyliowicz JS (1997) Denver international airport: lessons learned. McGraw-Hill, New York

14. Dessai S, Van der Sluijs JP (2007) Uncertainty and climate change adaptation - a scoping study. Copernicus Institute, Utrecht University, Utrecht

15. Dessai S, Van der Sluijs JP (2011) Modelling climate change impacts for adaptation assessments. In: Senn S, Dawid P, Christie M, Cliffen KA (eds) Simplicity, complexity and modelling. Wiley, Chichester, pp 83–102

16. Dewar JA (2002) Assumption-based planning: a tool for reducing avoidable surprises. Cambridge University Press, Cambridge

17. Dewar JA, Builder CH, Hix WM, Levin MH (1993) Assumption-based planning: a planning tool for very uncertain times. RAND Corporation, Santa Monica

18. Goodwin P, Wright G (2010) The limits of forecasting methods in anticipating rare events. Technol Forecast Soc Chang 77:355–368

19. Groot A, Rovisco A, Capela LT (2014) Showcasing practitioners' experiences. In: Lourenço TC, Rovisco A, Groot A, Nilsson C, Fussel HM, van Bree L, Street RB (eds) Adapting to an uncertain climate: lessons from practice. Springer, Berlin, pp 67–137

20. Groves DG, Lempert RJ (2007) A new analytic method for finding policy-relevant scenarios. Glob Environ Chang 17:73–85

21. Groves DG, Fischbach JR, Bloom E, Knopman D, Keefe R (2012) Adapting to a changing Colorado River: making future water deliveries more reliable through robust management strategies. RAND corporation, Santa Monica

22. Haasnoot M, Middelkoop H, Offermans A, van Beek E, van Deursen WPA (2012) Exploring pathways for sustainable water management in river deltas in a changing environment. Clim Chang 115(3–4):795–819.

23. Haasnoot M, Kwakkel JH, Walker WE, Ter Maat J (2013) Dynamic adaptive policy pathways: a new method for crafting robust decisions for a deeply uncertain world. Glob Environ Chang 23(2):485–498.

24. Haasnoot M, van Deursen WPA, Guillaume JHA, Kwakkel JH, van Beek E, Middelkoop H (2014) Fit for purpose? Building and evaluating a fast, integrated model for exploring water policy pathways. Environ Model Softw 60:99–120.

25. Hamarat C, Kwakkel JH, Pruyt E (2013) Adaptive robust design under deep uncertainty. Technol Forecast Soc Chang 80(3):408–418.

26. Hillier FS, Liebermans GJ (2001) Introduction to operations research. McGraw Hill

27. Holling CS (1973) Resilience and stability of ecological systems. Annu Rev Ecol Syst 4:1–23

28. Kabat P, Beekma J (1994) Water in the unsaturated zone. In: Ritzema H (ed) Drainage principles and applications, 2nd edn. International Institute for Land Reclamation and Improvement, Wageningen

29. Kahan JP, Botterman M, Cave J, Robinson N, Shoob R, Thomson R, Valeri L (2004) Cyber Trust and Crime Prevention: Gaining Insight from Three Different Futures. Prepared for Foresight Directorate, Office of Science and Technology, UK

30. KNMI (2006) Climate Change Scenarios 2006 for the Netherlands. KNMI de Bilt, Netherlands

31. Knol A, Slottje P, van der Sluijs JP, Lebret E (2010) The use of expert elicitation in environmental health impact assessment: a seven step procedure. Environ Heal 9(1):19

32. Kwadijk JCJ, Haasnoot M, Mulder JPM, Hoogvliet MMC, Jeuken ABM, van der Krogt RAA, van Oostrom NGC, Schelfhout HA, van Velzen EH, van Waveren H, de Wit MJM (2010) Using adaptation tipping points to prepare for climate change and sea level rise: a case study in the Netherlands. Wiley Interdiscip Rev Clim Chang 1(5):729–740.

33. Kwakkel JH, Walker WE, Marchau VAWJ (2010) Adaptive airport strategic planning. Eur J Transp Infrastruct Res 10(3):227–250

34. Kwakkel JH, Mens MJP, de Jong A, Wardekker JA, Thissen WAH, van der Sluijs JP (2011) Uncertainty Terminology. Knowledge for Climate. 30pp. http://goo.gl/H91ei

35. Lehner B, Döll P, Alcamo J, Henrichs T, Kaspar F (2006) Estimating the Impact of Global Change on Flood and Drought Risks in Europe: a continental integrated analysis. Clim Chang 75:273–299.

36. Lempert RJ (2002) A new decision sciences for complex systems. Proc Natl Acad Sci U S A 99:7309–7313

37. Lempert RJ, Popper S, Bankes S (2003) Shaping the next one hundred years: new methods for quantitative, long term policy analysis. RAND, Santa Monica

38. Lourenço TC, Rovisco A, Groot A, Nilsson C, Fussel HM, van Bree L, Street RB (eds) (2014a) Adapting to an uncertain climate: lessons from practice. Springer, Berlin

39. Lourenço TC, Rovisco A, Groot A (2014b) Making adaptation decisions under uncertainty: lessons from theory and practice. In: Lourenço TC, Rovisco A, Groot A, Nilsson C, Fussel HM, van Bree L, Street RB (eds)

Adapting to an uncertain climate: lessons from practice. Springer, Berlin, pp 139–162

40. Malatinszky Á, Ádám S, Saláta-Falusi E, Saláta D, Penksza K (2013) Planning management adapted to climate change effects in terrestrial wetlands and grasslands'. Int J Glob Warming 5(3):311–325

41. Marchau VAWJ, Walker WE, Van Duin R (2009) An adaptive approach to implementing innovative urban transport solutions. Transp Policy 15:405–412

42. Matrosov ES, Woords AM, Harou JJ (2013a) Robust decision making and info-gap decision theory for water resource system planning. J Hydrol 494:43–58

43. Matrosov ES, Padula S, Harou JJ (2013b) Selecting portfolios of water supply and demand management strategies under uncertainty - contrasting economic optimisation and 'robust decision making' approaches. Water Resour Manag 27:1123–1148.

44. Maxim L, Van der Sluijs JP (2011) Quality in environmental science for policy: assessing uncertainty as a component of policy analysis. Environ Sci Pol 14(4):482–492

45. McDaniel RR, Driebe DJ (eds.) (2005) Uncertainty and suprise in complex systems: questions on working the unexpected. Springer

46. Mens MJP, Klijn F (2015) The added value of system robustness analysis for flood risk management illustrated by a case on the IJssel River. Nat Hazards Earth Syst Sci 15:213–223.

47. Mens MJP, Klein F, De Bruijn KM, Van Beek E (2011) The meaning of system robustness for flood risk management. Environ Sci Pol 14(8):1121–1131

48. Mens M, Gilroy K, Wiliams D (this issue) Drought risk management under climate change uncertainty: defining critical system thresholds based on examples from the United States. Water Resources Management, this issue

49. Milly PCD, Betancourt J, Falkenmark M, Hirsch RM, Lettenmaier DP, Stouffer RJ (2008) Stationarity is dead: whither water management? Science 319(5863):573–574.

50. Mulder HM, Veldhuizen AA (2014) AGRICOM 2.01 Manual (in Dutch). Alterra, Wageningen, the Netherlands

51. Offermans A (2012) Perspectives Method: towards socially robust river management. University of Maastricht, Maastricht, the Netherlands.

52. Oosterbaan RJ (2001) SALTMOD: description of principles. User manual and examples of application. ILRI

53. PBL (2013) The effects of climate change in the Netherlands: 2012. PBL Netherlands Environmental Assessment Agency, The Hague

54. Popper S, Griffin J, Berrebi C, Light T, Min EY (2009) Natural Gas and Israel's energy future: a strategic analysis under conditions of deep uncertainty. RAND Corporation, Santa Monica

55. Quade ES (1982) Analysis for public decisions. Elsevier Science Publishing Co. Inc, New York

56. Rijnland (2009) Water management plan Rijnland 2010–2015 (In Dutch: Waterbeheerplan 2010–2015). Leiden, the Netherlands: Waterboard of Rijnland

57. Snelder M, van Zuylen HJ, Immers LH (2012) A framework for robustness analysis of road networks for short term variations in supply. Transp Res A Policy Pract 46:828–842.

58. Stelling J, Sauer U, Szallasi Z, Doyle FJ III, Doyle J (2004) Robustness of cellular functions. Cell 118:675–685.

59. Stichting Ontwikkeling De Venen (2010) "Gebiedsconvenant Groot Wilnis – Vinkeveen 2010–2020". Stichting Ontwikkeling De Venen, Utrecht

60. Swanson D, Barg S, Tyler S, Venema H, Tomar S, Bhadwal S, Nair S, Roy D, Drexhage J (2010) Seven tools for creating adaptive policies. Technol Forecast Soc Chang 77:924–939

61. ter Maat J, van Velzen E, van der Vat M (2013) Landelijke verkenning van effecten van maatregelpakketten voor de zoetwatervoorziening in Nederland. Deltares, Delft, the Netherlands

62. Urbonas BR, Roesner L (1993) Hydrologic design for urban drainage and flood control. In: Maidment DR (ed) Handbook of Hydrology. McGraw-Hill, New York, pp 28.01–28.52

63. Van Asselt MJA (2000) Perspectives on uncertainty and risk. Springer, Berlin

64. Van den Boogaard H, van Velzen E (2012) Empirical Relation between salt concentration in Rijnmond and the Rhine discharge. Deltares, Delft, The Netherlands

65. Van der Sluijs JP, Petersen AC, Janssen PHM, Risbey JS, Ravetz JR (2008) Exploring the quality of evidence for complex and contested policy decisions. Environ Res Lett 3:024008.

66. Van Geenhuizen M, Thissen WAH (2007) A framework for identifying and qualifying uncertainty in policy making: the case of intelligent transport systems. In: Van Geenhuizen M, Reggiani A, Rietvelt P (eds) Policy analysis of transport networks. Ashgate, Alsdershot

67. Van Geenhuizen M, Reggiani A, Rietveld P (2007) New trends in policymaking for transport and regional network integration. In: Van Geenhuizen M, Reggiani A, Rietvelt P (eds) Policy analysis of transport networks. Ashgate, Alsdershot

68. Van BL, Van der Sluijs JP (2014) Dealing with uncertainty in climate adaptation decision-making. In: Lourenço TC, Rovisco A, Groot A, Nilsson C, Fussel HM, van Bree L, Street RB (eds) Adapting to an uncertain climate: lessons from practice. Springer, Berlin, pp 17–40

69. Varum CA, Melo C (2010) Directions in scenario planning literature: a review of the past decades. Futures 42:355–369

70. Verhoeven J, Paulissen M, Ouboter M, van der Wielen S, Wegman R, Masselink L, Goosen H (2012) Klimaateffecten op de Natura 2000 moerascorridor: quick scan in het Groene Hart. Province of Zuid-Holland, The Hague

71. Walker B, Salt D (2006) Resilience thinking: sustaining ecosystems and people in a changing world. Island Press, Washington

72. Walker WE, Rahman SA, Cave J (2001) Adaptive policies, policy analysis, and policy-making. Eur J Oper Res 128(2):282–289

73. Walker W, Harremoes P, Rotmans J, van der Sluijs J, van Asselt M, Janssen P, Krayer von Krauss M (2003) Defining uncertainty: a conceptual basis for uncertainty management in model-based decision support. Integr Assess 4(1):5–18

74. Walker B, Holling CS, Carpenter SR, Kinzig A (2004) Resilience, adaptability and transformability in social-ecological systems. Ecol Soc 9(2):5–13

75. Walker WE, Haasnoot M, Kwakkel JH (2013) Adapt or perish: a review of planning approaches for adaptation under deep uncertainty. Sustain 2013(5):955–979.

76. Wardekker JA (2011) Climate change impact assessment and adaptation under uncertainty. PhD Thesis, Utrecht University. http://www.nusap.net/downloads/Wardekker_PhDdissertation_2011.pdf

77. Wardekker JA, de Jong A, Knoop JM, van der Sluijs JP (2010) Operationalising a resilience approach to adapting an urban delta to uncertain climate changes. Technol Forecast Soc Chang 77(6):987–998

78. Watt KEF, Craig PP (1986) System stability principles. Syst Res 3:191–201

79. Watts G, Von Christierson B, Hannaford J, Lonsdale K (2012) Testing the resilience of water supply systems to long droughts. J Hydrol 414–415(0), 255–267

80. Weaver CP, Lempert RJ, Brown C, Hall JW, Revell D, Sarewitz D (2013) Improving the contribution of climate model information to decision making: the value and demands of robust decision making frameworks. WIREs Clim Chang 4:39–60.

81. Wise RM, Fazey I, Stafford Smith M, Park SE, Eakin HC, Archer van Garderen ERM, Campbell B (2014) Reconceptualizing adaptation to climate change as part of pathways of change and response. Glob Environ Chang.

Chapter 7

An Integrated Framework for Assessment of Hybrid Water Supply Systems

Mukta Sapkota [1,*], Meenakshi Arora [1], Hector Malano [1], Magnus Moglia [2], Ashok Sharma [3], Biju George [4] and Francis Pamminger [5]

[1]Department of Infrastructure Engineering, Melbourne School of Engineering, University of Melbourne, Melbourne, VIC 3010, Australia

[2]Commonwealth Scientific and Industrial Research Organization (CSIRO), Land and Water, Clayton South, VIC 3169, Australia

[3]Institute of Sustainability and Innovation, Victoria University, Melbourne, VIC 3030, Australia

[4]International Centre for Agricultural Research in the Dry Areas, P.O. Box 2416, Cairo, Egypt

[5]Yarra Valley Water, Mitcham, VIC 3132, Australia

ABSTRACT

Urban water managers around the world are adopting decentralized water supply systems, often in combination with centralized systems. While increasing demand for water arising from population growth is one of the primary reasons for this increased adoption of alternative technologies, factors such as climate change, increased frequency of extreme weather events and rapid urbanization also contribute to an increased rate of adoption of these technologies. This combination of centralized-decentralized water systems approach is referred to as "hybrid water supply systems" and is based on the premise that the provision of alternative water sources at local scales can both extend the capacity of existing centralized water supply infrastructures, and improve resilience to variable climatic conditions. It is important to understand, however, that decentralized water production and reuse may change the flow and composition of wastewater and stormwater, thereby potentially also having negative impacts on its effectiveness and performance. This paper describes a framework to assess the interactions between decentralized water supply systems and existing centralized water servicing approaches using several analytical tools, including water balance modelling, contaminant balance modelling and multi-criteria decision analysis. The framework enables the evaluation of impacts due to change in quantity and quality of wastewater and stormwater on the existing centralized system arising from the implementation of hybrid water supply systems. The framework consists of two parts: (1) Physical system analysis for various potential scenarios and (2) Ranking of Scenarios. This paper includes the demonstration of the first part of the framework for an area of Melbourne, Australia by comparing centralized water supply scenario with a combination of centralized water supply and reuse of treated waste water supply scenario.

KEYWORDS

hybrid water supply systems; urban water cycle; evaluation framework; water balance; water reuse and recycling; resilience

1. INTRODUCTION

In most cities around the world, water services systems are mainly centralized. This is partly due to historical reasons, and in part due to the maturity of the technology and the opportunities for economics of scale. However, the lifespan and the planning horizon of central water infrastructures can be up to 100 years making the prediction of climate change, water availability, population growth as well as shrinkage and land use changes uncertain [1]. For instance, in some cities there are concerns about over-reliance on centralized systems in terms of their future sustainability for meeting a range of social, economic and environmental goals as well as for meeting the water demand because of the increasing growth and concentration of population in urban centers [2,3,4]. In some cities with shrinking populations as in Eastern Germany [5,6] and Japan [7] there are concerns about low flows in sewer pipe causing reduced critical traction forces causing sewer blockages [8] and ageing of water supply systems [1]. Centralized systems can also be vulnerable to occurrences of periodic droughts and the projected impacts of climate change [9]. Therefore, the case has been frequently made to augment the existing supply systems by integration of traditional and non-traditional supply sources [10]. This type of integration helps both in cases of increasing and decreasing population by making centralized organized water infrastructure more flexible and adaptable [1]. In the last 20 years, many alternative (non-traditional) water supply options like rainwater, stormwater and recycled water have emerged. The combination of such decentralized water supply options with centralized system which is defined as hybrid water supply systems by Sapkota, *et al.* [4], have displayed potential to meet the increasing water demand [11,12]. It is argued that this type of hybrid approach can thus offer flexible solutions, wherever certain thresholds of population density are exceeded [13,14]. These approaches help reduce the volume of water imported to cities, and in some configurations can also reduce the volumes of wastewater and stormwater discharged into the environment [15]. On the other hand, the use of these systems may have some detrimental impacts on the centralized water infrastructure, particularly on the sewage network and stormwater drains. The use of decentralized water supply systems changes both the wastewater and stormwater flow regimes and contaminants' composition [16,17,18]. For instance, as water saving and recycling are encouraged through grey water reuse, dual piping systems, and sewer mining, the quantity of wastewater may decrease leading to an increase in its concentration of contaminants [19]. High-concentration wastewater has been known to cause sewer problems such as sewer blockage, odor and corrosion [20]. Thus, to ensure the smooth and effective implementation of hybrid water supply systems, it is necessary to evaluate their impacts on the existing wastewater and stormwater system. Such an evaluation requires a comprehensive methodology to enable the evaluation of various subsystems and their interactions to contribute to the overall performance of hybrid systems [4,21].

In most previous studies, integrated assessment of supply options has been carried out using narrow and opposing options (an either/or approach). Assessing one water supply alternative against another for a number of alternative systems has typically excluded hybrid options for partial contributions from multiple alternatives together [22]. For example Cook,*et al.* [23] analyzed seven alternatives and Abrishamchi, *et al.* [24] analyzed eight systems, however, these methodologies exclude options of combined systems and their potential benefits. In fact, until now the perceived benefits of combined centralized and decentralized approaches has only been discussed predominantly on a theoretical and ideological basis [2,25,26]. There is still a question of whether a mixture of a decentralized and centralized approach can combine the benefits of both systems. To answer this question of finding appropriate balance of centralized and decentralized servicing options for urban water management, we must rely on a robust understanding based on structured modelling approaches and reliable data.

At present, there is very scant literature that present assessment frameworks [14,21,22,27] and empirically based evidence [28,29,30] leading to an objective assessment of these hybrid systems. With few exceptions, none of the previously published methodologies evaluate the impacts of hybrid systems in terms of the interaction between centralized and decentralized infrastructures. Some of the existing frameworks evaluate the environmental impacts of hybrid systems, however, they do not evaluate the impacts arising from the implementation of hybrid systems on the quantity and quality of wastewater and stormwater on the existing centralized systems [31]. Hence there is a need of comprehensive methodology for assessing conjointly the full range of technological alternatives available. The closest methodology for assessing the introduction of hybrid water system to date is presented by Sapkota, *et al.* [21]. However, this is a basic theoretical methodology that requires further empirical study to validate it via a case study application. This framework is described in this study together with an expanded description of the evaluation of hybrid water supply systems and some modifications of the methodology. This study demonstrates the systematic implementation of a subset of this framework focused on the physical aspect of the system. The analysis is based on analytical approaches to assist in understanding the interaction between centralized and decentralized systems.

2. CONCEPTUAL FRAMEWORK

2.1. Urban Water Cycle

Figure 1 presents the stock-and-flow representation of a hybrid water supply system. It shows the main components of the water cycle, pathways and alternative supply options considered in this study. Demand is mainly governed by changes in population and variability in climate. To meet this demand, water can be supplied on a fit-for-purpose basis from various decentralized options (greywater, wastewater, stormwater and roof water) along with centralized supply. Unused wastewater and greywater flows to the receiving waters via a wastewater treatment plants while excess roof and grounds runoff flows to the receiving waters. In this model, water supply, wastewater and stormwater are considered within a single integrated framework to highlight their interactions. In addition, a broad range of non-conventional water supply technologies such as rainwater tanks, stormwater harvesting, and greywater reuse are considered as part of the system to analyze their impacts on the urban water cycle components. Because of the interconnectedness of the hybrid system, changes in one component of the system can have impacts on other components. The framework also considers the flow of contaminants throughout the urban area.

2.2. System Boundary

Both centralised and decentralised water supply systems are considered in this study. Centralised techniques are referred to here as those applications that favour city-wide solutions, including wide-area distribution networks, large-scale water storage facilities and the development of large scale resources.

Decentralised options are defined as those being applicable at development or individual household level [9]. For the purpose of modelling, a range of sub-systems have been considered for the study: water use in residential, industrial, commercial and public areas; central water supply; urban waterways and stormwater; sewerage and wastewater treatment; and provision of alternative water sources at different scales (development, neighbourhood or allotment). The allotment/household scale herein represents a building and associated indoor and outdoor usages as well as paved and pervious areas such as pathways, driveways and gardens. The neighbourhood comprises a number of identical unit blocks as well as roads and public open space. The development/subdivision scale represents the grouping of one or more neighbourhoods that may or may not have the same land use or water servicing approach. Rainwater tanks and grey water recycling are considered at household level while stormwater harvesting and recycled wastewater are assumed to be implemented at development scale. However, rainwater usage can also be considered at development scale through the use of communal rainwater tanks [32].

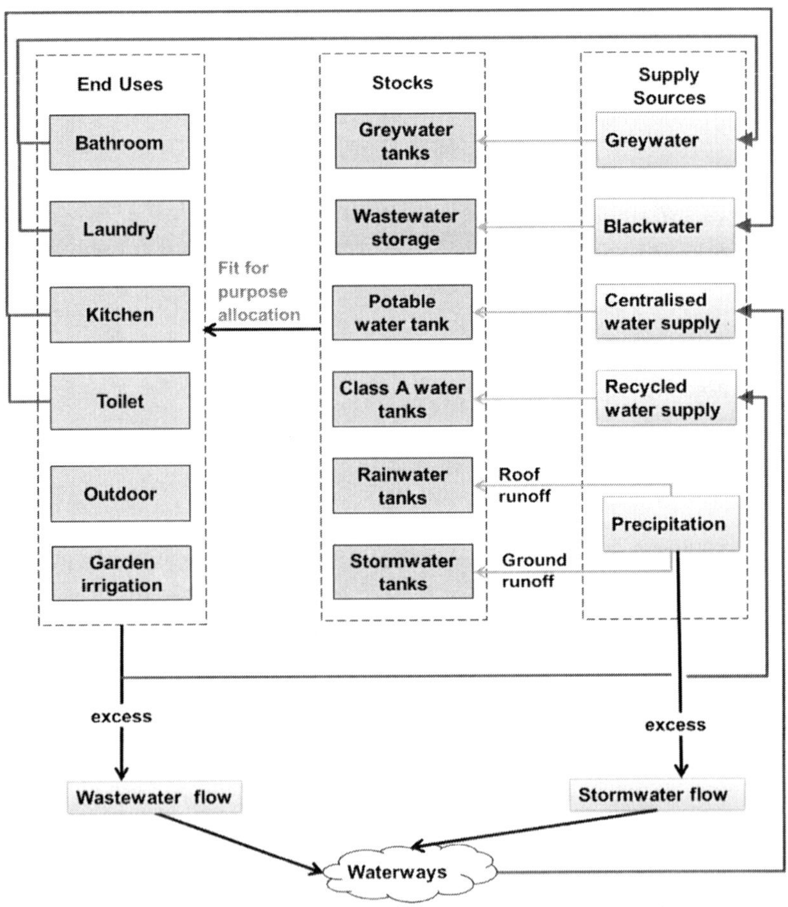

Figure 1. Stock-and-flow diagram of hybrid WSS (Water Supply System).

2.3. Variables

Water flow rates, volumes and contaminants are considered in the analysis and the specific variables included in the study are water (potable water, stormwater, rainwater, greywater and wastewater) flow rates and volumes along with contaminant loads (Total Nitrogen, Total Phosphorous, Total Suspended Solids, BOD and COD). TSS, TP, TN, BOD and COD are selected to represent the contaminant loads as they are most significant measures of pollution in both runoff and wastewater [33]. Flow rates are important to assess the infrastructure capacity such as the maximum flow that water, stormwater and sewage pipes can handle. Volume is important in terms of designing water supply reservoirs, and water and wastewater treatment plant capacity.

3. FRAMEWORK TO EVALUATE HYBRID WATER SUPPLY SYSTEMS

In this study we adopted a generalized framework (Figure 2) from Sapkota, *et al.* [21] that considers the varying nature of urban developments and forms. The framework is supported by several models and tools including water balance modelling, contaminant balance modelling, multi-criteria decision analysis (MCDA), and future change analysis. The framework has been divided into two parts: the analysis of the physical system and the ranking method. In this study, the physical system analysis is applied to a case study for the purpose of illustration consisting of two typical scenarios. A ranking methodology is proposed without an actual application which would require a larger set of scenarios.

3.1. Physical System Analysis

3.1.1. Understanding the Local Conditions and Current System Capacity

This component of the framework explores the local conditions such as local climate, geology, development pattern, future population projection and water demand, water resource availability, environmental flow requirement and wastewater and stormwater disposal limits, and system capacity quantities such as capacity of the water supply, sewage and stormwater drainage networks; water, stormwater and wastewater treatment plants. This task involves reviewing of relevant literature including reports from government authorities, policy and guidelines. An essential part of this step is to define the system boundary. Physical boundaries are selected to encompass the area of interest (urbanized area at development scale) and also include elements such as centralized water supply, discharge points, treatment plants and receiving waters.

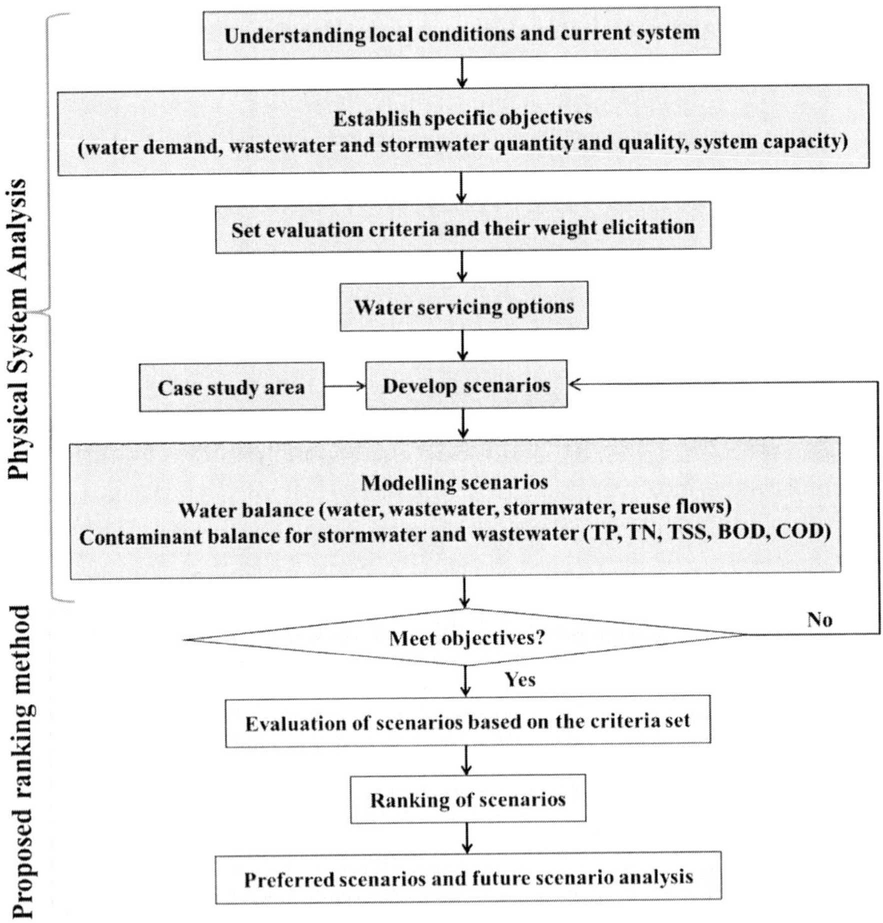

Figure 2. Framework to evaluate hybrid WSS.

3.1.2. Establishing Specific Objectives

Specific objectives must be set to meet water supply and demand, quantity and quality of wastewater and stormwater discharge and system capacity. These characteristics are site specific and are based on the local information and systems capacity, present water demand profile, current water supply, sewage and drainage profiles, recycled water use profile and future water demand profile and rainfall pattern.

3.1.3. Setting the Evaluation Criteria and Their Weight Elicitation

The study presented herein considers mainly a set of quantitative criteria that reflects the main objective of the study which is to assess the impact of decentralized systems on existing centralized infrastructures. These criteria were set in consultation with Victorian water utilities through series of workshops and personal interviews. The assessment criteria are set keeping in mind that the study is confined to the physical aspects of the system. The criteria set for the evaluation of the performance of hybrid water supply systems include:

1) Reduction in potable water demand from centralized WSS

2) Reduction in wastewater discharges both flow rate and volumes

3) Reduction in contaminant loads of wastewater flow

4) Reduction in stormwater flows both intensity and volumes

5) Reduction contaminant loads from stormwater to receiving water

6) Improvement of supply reliability of fit for purpose water

The multi-criteria decision analysis (MCDA) in this framework requires the definition of weights. The elicitation of meaningful weights for these criteria is very important to reflect their relative importance. For this purpose, a questionnaire survey has been conducted among water professionals from local water utilities, research organizations, universities, consultancies and water associations to determine the factors that need to be considered in the evaluation criteria and the relative importance (weight) of each criterion. The survey results from 37 respondents provided almost similar weights to each criterion varying between 0.18 and 0.15, indicating that all criteria have similar importance. Moreover, all respondents believe that none of the criteria are unimportant which demonstrates that the criteria used in the study are robust. It is widely acknowledged that the assessment of criteria weighting is a crucial step in multi-criteria decision analysis [34]. While providing these weights for MCDA, the study framework does recognize the importance of conducting a sensitivity analysis to examine the relative impacts on outputs of possible variations in weighting. However, this component is not demonstrated in this paper as the MCDA analysis is not incorporated in the case study application.

3.1.4. Water Supply Servicing Options

Identifying the feasible water supply options at the selected site is an important step. Options can vary based on allotment, neighborhood or development scale and availability of alternate sources. Examples of various alternative water supply options based on a fit for purpose are greywater collection and reuse, rainwater harvesting and reuse, stormwater harvesting and reuse, and sewer mining.

3.1.5. Developing Scenarios

In this stage, various hybrid (combination of centralized and different decentralized) water supply scenarios are developed in consultation with the water utilities that involve combinations of recycled water via third pipe, treated greywater, rainwater tanks and stormwater tanks. These are based on the combination of the selected water supply servicing options, infrastructure provision and chosen technologies. These scenarios are developed based on the current local/state Government policy and compliance with existing water legislation. For example, a hybrid water supply scenario may include potable water supply from centralized water supply system for kitchen, bath and laundry, while non-potable supply can be sourced from greywater reuse for toilet flushing and garden irrigation.

3.1.6. Analysis of Scenarios

A water balance analysis provides the volume and flow rate of potable water, stormwater and wastewater along with flow quantities of alternative resources based on defined end uses and size of storages for rain tanks, grey water, wastewater and stormwater reuse. There are various models available for water balance analysis such as Aquacycle [35], Hydro Planner [36], Krakatoa [37], UrbanCycle [38], UVQ [39], WaterCress [40], Urban Developer model [41] and City Water Balance [33]. The UVQ (Urban Volume and Quality) model [39,42] has been selected to carry out the water and contaminant balance analysis in this study due to its capacity for system integration, and options for consideration of varying spatial scale and water quality which are main requirements of the present integrated system analysis approach. The water and contaminant balance results from this analysis can then be used as one of the major components for evaluation of the hybrid water supply systems. In addition, supply reliability of fit-for-purpose water, defined as the percentage of demand met from the combination of alternative water supply storages over the modeling period will be used for the evaluation of the hybrid water supply system. Supply reliability is estimated based on the daily water balance provided by the UVQ model.

3.2. Proposed Ranking Method

3.2.1. Scenario Evaluation

In this segment of the framework, water and contaminant balance outcomes need to be evaluated against the specific objectives set out for water demand and system capacity. Scenarios found to meet these development specific objectives will be considered for Multi-criteria Decision analysis later.

3.2.2. Ranking of Scenarios

Several empirical and analytical approaches such as cost benefit analysis, life cycle analysis, community cost incorporating life cycle and environmental cost, and various multi-criteria assessment methods can be employed to provide an integrated assessment of hybrid water supply systems performance [22,31]. Of the various approaches, multi-criteria decision analysis (MCDA) is a technique that explicitly considers the multiple criteria in a decision making environment that aims to improve the transparency, auditability and analytical rigor of these decisions [43]. This approach has been widely utilized for urban water system analysis in several studies [24,44,45,46,47]. MCDA evaluates and ranks the finite number of decision options based on the weights of a finite set of criteria [45]. There are several MCDA methods used for water resources management problems. The most commonly applied methods includes fuzzy set analysis, Compromise Programming (CP), Analytic Hierarchy Process (AHP), Elimination et Choice Translating Reality (ELECTRE) and Preference Ranking and Organization Method for Enrichment Evaluation (PROMETHEE) [44]. These methods differ from each other in the methodology they use, the quality and quantity of the extra information they need, the sensitivity tools they offer, their ease of application, the mathematical properties they verify and availability of software [48]. The PROMETHEE method based on a pair-wise comparison [49,50] of options is proposed for this study because of its transparent computational procedure and the comparatively low time and effort needed by the Decision Maker to reach a conclusion [51]. The D-Sight Software [52,53] based on PROMETHEE will used in this study to rank alternative scenarios.

Several criteria along with their weights from Section 3.1.3 can be considered to evaluate the scenarios. Further, preference functions which signify the relative importance of one alternative over another with respect to the adopted evaluation criterion [51] needs to be assessed based on the questionnaire survey among water resource managers, stormwater planners, water supply planners, sewer planners and integrated urban water managers. Scenarios can be ranked based on the weights and preference function of the evaluation criteria; and outputs from hybrid water supply scenario analysis.

3.2.3. Preferred Set of Scenarios and Future Scenario Analysis

The preferred sets of scenarios need to be selected from the ranking of scenarios. These scenarios can then be modelled to determine the impacts of climatic and demographic changes. Such an approach helps in identifying the hybrid water supply scenario that performs best under a range of conditions. This analysis assists in identifying a resilient system. In this study, a resilient system is defined as a system that is able to absorb disturbances and still remain within the same state, reorganize itself while undergoing changes so as to still retain essentially the same function, and build and increase the capacity for adaptation [54]. The concept of a resilient system has been adopted considering its suitability for water supply and wastewater infrastructures [55].

As discussed earlier, this study demonstrates a part of the proposed framework. Specifically, a subset of this evaluation framework (indicated in blue color in Figure 2) is applied in the Northern Growth Area (NGA) of Melbourne (Figure 3), Australia as presented in the next section.

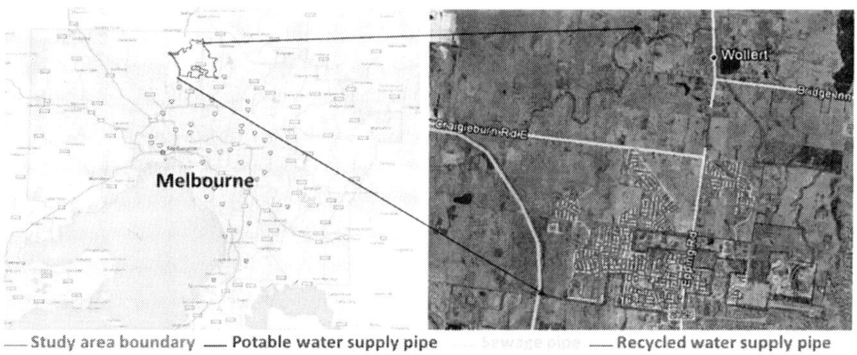

Figure 3. Location map of study area.

4. CASE STUDY APPLICATION

The Northern Growth Area (NGA) of Melbourne, Australia consisting of the urban development of Aurora, Epping North East, Quarry Hill and Wollert, is located approximately 20 km north from the Melbourne CBD. It is supplied with Class A recycled water from the Aurora wastewater treatment plant via third pipe for non-potable use which includes toilet flushing, garden watering, car washing and irrigation of public open space.

This section presents the systematic implementation of the methodology developed in this study and the analysis approach designed to understand the impacts of hybrid water supply systems.

The study area covers 2257 hectares of land. Based on the Google map accessed on 27 August 2012, there were 3455 residential households, a sport complex and a shopping complex. For modelling purpose, a residential area of 450 m^2 with roof area of 300 m^2, paved area of 60 m^2 and garden area of 90 m^2 was considered [56]. In addition, this study considers a household occupancy rate of 2.86 people per household [57]. Table 1 presents indoor water demand for residence provided by the local water utility Yarra Valley Water (YVW). Water use for toilet in commercial lots has been calculated assuming the equivalent of 180 users per day based on the technical report by Sharma, *et al.* [58]. Garden irrigation demand is based on soil moisture level and is calculated within the model.

Table 1. Indoor water demand for residential purposes.

Use	Demand (L/Cap/Day)
Toilet	21.6
Bathroom	62.5
Laundry	26.8
Kitchen	22.7

Climate data was adopted from Bureau of Meteorology, Australia records. A period of 34 years (1980–2013) was used for the water balance analysis of the area which has an annual average rainfall of 642 mm and pan evaporation of 1198 mm. GIS maps for potable/recycled water and sewage infrastructures were collected from YVW and stormwater drainage from Melbourne water. These maps provide the layout of the water supply and wastewater/stormwater networks together with their flow capacity. This information helped set the limits of the feasible ranges of water, wastewater and stormwater flow rate and contaminant concentration.

The water supply options assessed for the study area include centralized supply (conventional reticulated system), a third pipe system supplying treated wastewater or reclaimed water, rainwater tanks and greywater systems. Based on the most common integrated water management approaches adopted globally, the alternative water supply options includes rainwater, stormwater, wastewater and greywater reuse [31]. Current practices in Melbourne, Australia have adopted similar approaches [59].

Two typical scenarios are compared to illustrate the application of this framework. These scenarios include:

1) IScenario 1: This conventional scenario considers only a centralized water supply system scenario where the entire water demand is met by potable water. Please refer to the schematic diagram of the scenario in Figure A1. This scenario represents the conventional water supply system.

2) IIScenario 2: The centralized system is combined with treated recycled water (Class A water) via 3rd pipe. This scenario represents a typical scenario that illustrates the effects of a decentralized water supply option on the centralized water infrastructure. In addition, this case reflects the existing water supply system in the area. In this scenario (as shown inFigure A2), wastewater is collected at development level and distributed through a dual reticulation system for toilet flushing and garden irrigation after treatment. The remaining water demand is met from potable water supply. The implementation of this alternative system must always consider the use of suitable wastewater technology and its management to prevent any contamination risk from the recycled water. This aspect, however, is not dealt within the scope of this paper.

As discussed earlier the water and contaminant balance analysis is conducted using the UVQ (Urban Volume and Quality) model [39]. The water balance analysis estimates volume and flow rate of potable water, stormwater and wastewater along with flow quantities of wastewater reuse. The UVQ function enables the estimation of the contaminants balance and loads including TSS, TP, TN, BOD and COD. Here, contaminant balance estimates are based on the water volumes calculated in the water balance and user specified concentrations, loads and performance criteria. Removal percentage of specified contaminants with the use of different wastewater and stormwater technologies are obtained from the literature [60] and from data available from Yarra Valley Water (YVW), Victoria. The removal efficiency for treated waste water is calculated based on the actual data from Yarra Valley Water for raw effluent going to the treatment plant and Class A recycled water. The measured quality of Class A recycled water is aligned with water quality guidelines provided by EPA (Environmental Protection Agency), Victoria as shown in Table A1. The estimated values of

TN, TP, TSS, BOD and COD are 98.1%, 99.32%, 99.62%, 99.61% and 99.61%, respectively.

The model was run over a 34-year time series of climate data at daily time step to simulate seasonal and annual climate variability. For Scenario 2, a recycled water tank size of 1.0 mL was adopted consistent with the actual recycled water tank size used in the area. The supply reliability of the tank found that 96% of non-potable water demand (garden water and toilet water in this case) is met from the alternative water supply storage (treated wastewater) over the modeling period while the rest is provided from the centralized system. This reliability, which depends on the size of wastewater storage tank, is calculated based on the daily water balance. Treated wastewater could not meet the full demand despite the annual supply of wastewater being greater than the total annual demand for toilets and gardens because sufficient wastewater cannot be stored and given the seasonal variation in garden demand. Hence, increasing the size of recycled water tank can increase the system reliability. For instance, in this particular scenario, using a 1.1 mL tank would have met all the required demand. The water and contaminant balance summary along with the supply reliability of the two scenarios is presented inTable A2. The removal efficiency for treated wastewater is calculated based on the data from Yarra Valley Water for raw effluent going into the treatment plant and Class A recycled water. Figure 4 below presents the change in potable water flow, wastewater flow, wastewater contaminants and supply reliability.

Scenario 2 requires 16.97% less potable water and generates 24.64% less wastewater compared to Scenario 1. However, Scenario 2 results in increased contaminant concentration in wastewater by almost 32%. This is because the residual contaminants from decentralized local wastewater treatment are assumed to be disposed of to the main sewer. However, all the contaminant concentrations fall within the range of contaminant concentration for typical untreated domestic wastewater (Table A3) based on the existing literature [61,62,63] except for total phosphorous which is slightly higher. Based on the nature of decentralized wastewater treatment process, an appropriate sludge management approach needs to be adopted.

This study models two scenarios to estimate water and contaminant balance to demonstrate the application of the proposed framework. The baseline setting of the two scenarios was found to meet the specific objectives related to water demand and system capacity. To evaluate the impacts of changed demographics and climatic conditions, the two scenarios are subjected to the following perturbations:

1) ICondition 1—Scenarios are run with 10% increase in population

2) IICondition 2—Scenarios are run with changed climatic condition at a daily time scale. Climate data (Precipitation and Temperature) from 2030–2055 based on the RCP (Representative Concentration Pathway) 8.5 high emission scenarios are chosen for illustration purpose. RCP 8.5 is based on rising radiative forcing pathways leading to 8.5 W/m^2 (~1370 ppm CO_2 eq) by 2100 [64]. This scenario represents a High RCP characterized by increasing greenhouse gas emissions over time that lead to high greenhouse gas concentration levels [64]. A high RCP scenario was chosen for this analysis to represent the worst-case future scenario. Detail information of calculations related to greenhouse gas concentration is adopted from Meinshausen *et al.* (2011) [65]. In this scenario, average and maximum temperature increase is predicted to be 10% and 7%, respectively. Maximum precipitation increases by 0.3%, mean precipitation decrease by 1.1% and the number of wet days (precipitation ≥ 1 mm/day) decreases by 4.6%.

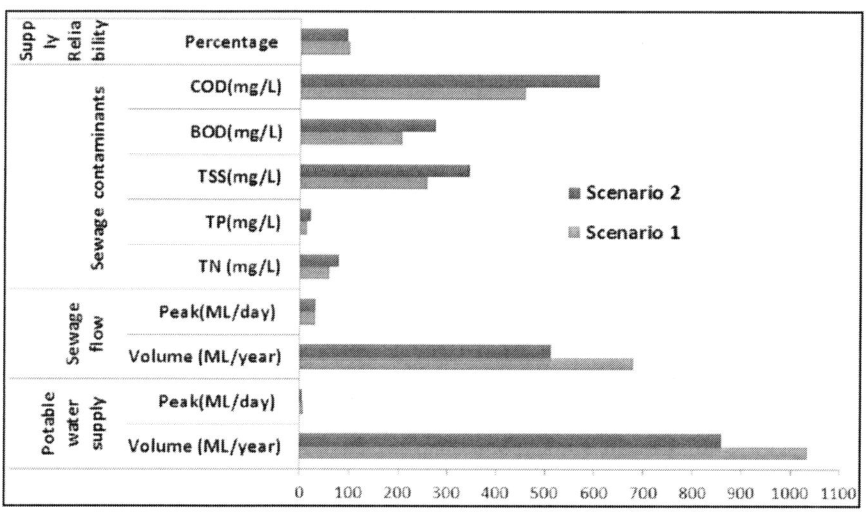

Figure 4. Average water output for the two scenarios for the selected criteria.

A summary of the water and contaminant balance for these scenarios and changed conditions is provided in Table A4. The comparison of scenarios 1 and 2 based on 2012 population level (Figure 4) shows a change in potable water and sewage flow. The change in potable water and wastewater variables in response to demographic and climatic changes is shown in Figure 5. Scenario 1 is taken as base scenario and assigned a value of one for variables set to 2012 levels. For other scenarios, variables are taken as a ratio between the variables of the respective scenario and the base Scenario 1 for 2012 population level.

When comparing the two scenarios for the changed conditions, it is found, as expected, that population changed conditions have major impacts on potable water use. Further, it is observed that the use of treated wastewater along with centralized supply can provide the extra water needed for the area with less change in daily average peak potable water supply flow. 3.97% decrease in potable water supply peak is observed for Scenario 2 compared to 14.90% increase in Scenario 1 for the increased population level. For the changed climatic condition, potable water supply volume is found to increase by 3.81%. This is primarily due to the additional use of potable water for garden irrigation as average precipitation is lower in the RCP 8.5 climate scenario. The wastewater flow volume increases by 5.84% with the increase in population, however, a negligible increase in peak flow is observed. The peak wastewater flow is mainly governed by wet flow which includes sanitary flow and a portion of storm water flow. For climate change, the volume of wastewater increases by 4.76% because there is an increase in the numbers of higher intensity rainfall events in the RCP 8.5 climate scenarios. No significant difference is observed in the maximum (peak) wastewater flow as there is only 0.3% increase of maximum precipitation accompanied by an increase in average temperature of 7.0%.

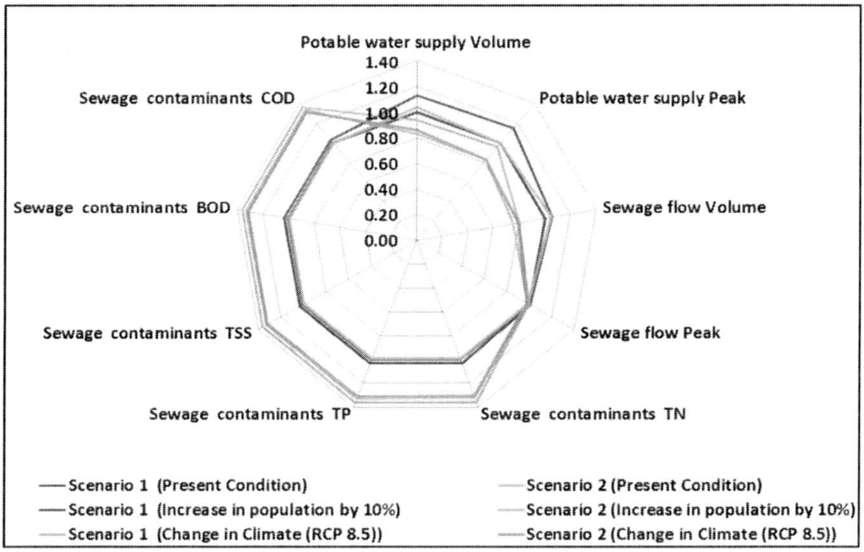

Figure 5. Spider diagram comparing two scenarios in various conditions.

Figure 6 shows the box plot of daily potable water use for the two scenarios under climate and population change conditions. The plot suggests that scenario 2 under all conditions has a narrow range of variation compared to scenario 1. In addition, scenario 2 has a lower-volume of maximum daily potable water supply in all the cases which makes it more resilient under the changed condition relative to scenario 1.

Figure 7 shows the box plot of daily sewage outflow for the two scenarios for changed climate and perturbated population levels. The overall box plot shows a large variation in high flows. It is because high flows not only include dry weather (sanitary) flow but also groundwater non-rainfall dependent infiltration, and rainfall dependent inflow. Despite the existence of a separate collection system, it is virtually impossible to isolate sewage system from storm rainfall [66]. As shown in Table A2 and Table A4, climate change has no significant effects on the contaminant concentration of wastewater. These results also show that when population increases, contaminant concentration in the wastewater increases in the centralized system. While comparing Scenario 2 under population increase condition with Scenario 2 of 2012, the wastewater concentration is observed to increase. This can be ascribed to an increase in population and an increase in the reuse of wastewater.

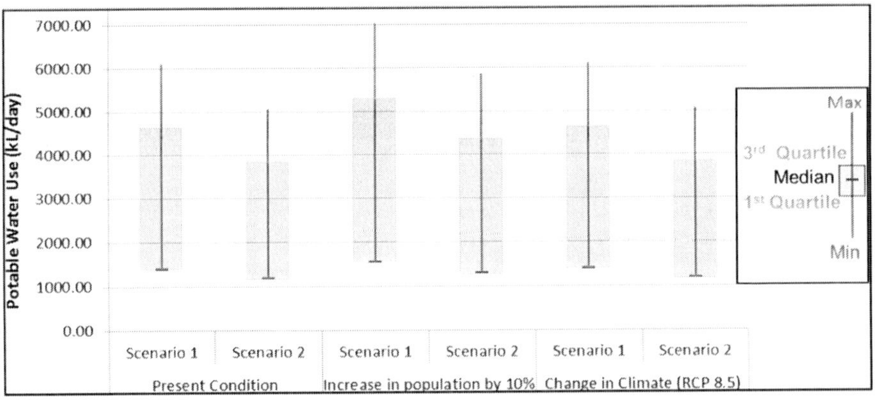

Figure 6. Box plot of potable water use under two scenarios for changes in population and climate.

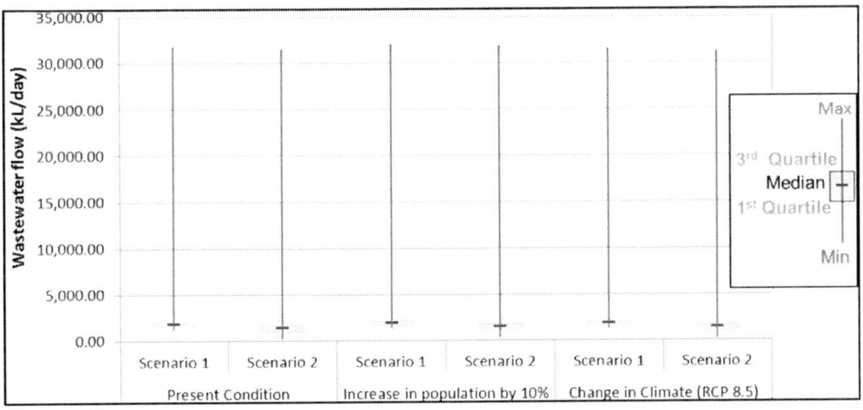

Figure 7. Box plot of wastewater flow under two scenarios for changes in population and climate.

5. DISCUSSION

This study provides a comprehensive framework to evaluate the impacts of hybrid centralized and decentralized urban water supply systems which are increasingly being considered as a viable option for meeting increasing water demand under resource scarcity and the use of a diversity of supply options. In this study, we have analytically demonstrated that there are likely interactive impacts from the implementation of hybrid WSS that can have implications on the operation and performance of existing centralized infrastructure including possible impacts on changes in wastewater and stormwater quantity and quality.

Proposed framework is illustrated through the applications on the Northern Growth Area, a recent urban development in Melbourne, Australia. In this case study, a combination of a typical centralized water supply system with treated waste water system is assessed to estimate changes in water and contaminant balance in relation to a centralized only system. The results from water and contaminant balance modelling show that the use of hybrid water supply systems reduces the potable water demand significantly (1036 to 860 mL/year). Further, wastewater flow is also reduced from 678 mL/ year to 511 mL/year. Conversely, wastewater contaminant concentration for TN, TP, TSS, BOD and COD increases for the hybrid scenario by almost 32%.

In addition, the two water supply scenarios were evaluated for changes in demographics and population. When compared to the centralized-only system, the hybrid water supply scenario is found to be more resilient in terms of adjusting to the changing population and climatic conditions as less variation is observed in potable water demand. In this way, the case study findings help to demonstrate that the framework is suitable to assess the impacts of hybrid water supply systems.

In future, this study will consider wider ranges of scenarios and assess them using multi-criteria decision analysis as a tool to rank their performance. The water and contaminant balance resulting from the analysis can be used as one of the main elements in the evaluation of hybrid water supply systems.

This study, however, has some limitations that need to be taken into account in future research. First, it has not considered social behavior and local acceptance of hybrid water supply systems. A more qualitative study is required to identify the perception of people which is beyond the scope of this study [67]. Second, this research does not take into account the energy usage and economic aspects of hybrid water supply system. However, the evaluation framework presented in this study is sufficiently flexible to incorporate the socio-economic dimensions if so required. Therefore, the current framework should only be used to assess the physical interaction of hybrid WSS with centralized water system in terms of flow and variations in contaminant loads. In the context of public health protection, it is assumed that the application of this framework must always consider the relevant legal and health regulations.

6. CONCLUSIONS

This study presents an evaluation framework coupled with water and contaminant balance modelling, future scenario analysis and multi-criteria decision analysis to assess the interaction between centralized and decentralized water systems. Using this framework, hybrid water supply scenario is analyzed for its impacts on potable water, wastewater and stormwater flows. Case study application demonstrates that the use of alternative water supply options combined with centralized water supply system can significantly reduce potable water demand and alter wastewater flow and contaminant concentration. For instance, hybrid water supply scenario uses 16.97% less potable water and generates 24.64% less waste water flow compared to conventional centralized scenario. Conversely, wastewater concentration is increased by 32%. The study also shows that use of the hybrid water supply scenario reduces the variability in potable water use (1.16–5.05 (mL/day)) compared to conventional system (1.4–6.11 (mL/day)). These scenarios are also evaluated for changes in demographics and population. Results show population change has more significant effect on potable water use compared to climate change. For Scenario 1, potable water supply volume and peak is changed by 13.31% and 14.90% for increased population condition. For climate change condition, only potable water supply volume was increased by 3.81%. Wastewater flow volume is increased by 5.84% and 4.76% respectively, for changed climatic and demographic conditions respectively compared to Scenario 1 of 2012. Hybrid scenario is found to provide extra water needed for the area with relatively less change in daily average peak potable water supply under the changed condition.

Hence, this framework contributes in understanding and implementing hybrid water supply system in an objective manner which is crucial in the development of more sustainable and resilient urban infrastructure system. In the context of climate change and population growth this study framework adds further value to develop resilient urban water infrastructure systems. In this way, the case study findings helped to demonstrate that the framework is suitable to assess hybrid water supply systems. Moreover, this framework contributes in understanding and implementing hybrid water supply system in an objective manner which is crucial in the development of more sustainable and resilient urban infrastructure system. In the context of climate change and population growth this study framework adds further value to develop resilient urban water infrastructure systems. The use of such framework can help water managers to make informed decisions involving the combination of decentralized and centralized systems under a variety of population and climate scenarios.

ACKNOWLEDGMENTS

The authors would like to thank Yarra Valley Water (Melbourne, Australia), Melbourne Water (Melbourne, Australia) and Bureau of Meteorology (Melbourne, Australia) for providing the data. Mukta Sapkota is conducting this PhD study with funding support from the Australian Government and the Melbourne University.

AUTHOR CONTRIBUTIONS

All authors were involved in designing and discussing the study. Mukta Sapkota drafted and finalized the manuscript. She executed the model. Meenakshi Arora, Hector Malano, Magnus Moglia, Ashok Sharma, Biju George and Francis Pamminger supervised the research. All authors discussed the results and implications and commented on the manuscripts at all the stages.

CONFLICTS OF INTEREST

The authors declare no conflict of interest.

REFERENCES

1. Sitzenfrei, R.; Rauch, W. Investigating transitions of centralized water infrastructure to decentralized solutions—An integrated approach. *Procedia Eng.* **2014**, *70*, 1549–1557.

2. Marlow, D.R.; Moglia, M.; Cook, S.; Beale, D.J. Towards sustainable urban water management: A critical reassessment.*Water Res.* **2013**, *47*, 7150–7161.

3. Lee, S.; Yigitcanlar, T.; Egodawatta, P.K.; Goonetilleke, A. Sustainable water provision: Challenges, alternative strategies and sources in the era of climate change. In *Sustainable Urban and Regional Infrastructure Development: Technologies, Applications and Management*; Yigitcanlar, T., Ed.; IGI Global: Hershey, PA, USA, 2010; pp. 17–30.

4. Sapkota, M.; Arora, M.; Malano, H.; Moglia, M.; Sharma, A.; George, B.; Pamminger, F. An overview of hybrid water supply systems in the context of urban water management: Challenges and opportunities. *Water* **2015**, *7*, 153–174.

5. Schramm, E.; Felmeden, J. Towards more resilient water infrastructures. In *Resilient Cities 2: Cities and Adaptation to Climate Change—Proceedings of the Global Forum 2011*; Otto-Zimmermann, K., Ed.; Springer: Berlin, Germany, 2012; Volume 2, pp. 177–186.

6. Großmann, K.; Bontje, M.; Haase, A.; Mykhnenko, V. Shrinking cities: Notes for the further research agenda. *Cities***2013**, *35*, 221–225.

7. Daigger, G.T. Evolving urban water and residuals management paradigms: Water reclamation and reuse, decentralization, and resource recovery. *Water Environ. Res.* **2009**, *81*, 809–823.

8. Swamee, P.K.; Bhargava, R.; Sharma, A. Noncircular sewer design. *J. Environ. Eng.* **1987**, *113*, 824–833.

9. Sharma, A.K.; Tjandraatmadja, G.; Cook, S.; Gardner, T. Decentralised systems—Definition and drivers in the current context. *Water Sci. Technol.* **2013**, *67*, 2091–2101.

10. Skinner, R. Adaptation to climate change in melbourne: Changing the fundamental planning assumptions. In International Adaptation Forum on Climate Change Impacts on Water Supply, Washington, DC, USA, 27 January 2010; p. 24.

11. Rozos, E.; Makropoulos, C. Assessing the combined benefits of water recycling technologies by modelling the total urban water cycle. *Urban Water* **2012**, *9*, 1–10.

12. Gikas, P.; Tchobanoglous, G. The role of satellite and decentralized strategies in water resources management. *J. Environ. Manag.* **2009**, *90*, 144–152.

13. Bieker, S.; Cornel, P.; Wagner, M. Semicentralised supply and treatment systems: Integrated infrastructure solutions for fast growing urban areas. *Water Sci. Technol.* **2010**, *61*, 2905–2913.

14. Weber, B.; Cornel, P.; Wagner, M. Semi-centralised supply and treatment systems for (fast growing) urban areas. *Water Sci. Technol.* **2007**, *55*, 349–356.

15. Coombes, P.J.; Kuczera, G. Integrated urban water cycle management: Moving towards system understanding. In Proceedings of the 2nd National Conference on Water Sensitive Urban Design, Brisbane, Australia, 2–4 September 2002.

16. Butler, D.; Makropoulos, C. *Water Related Infrastructure for Sustainable Communities*; Environment Agency: Bristol, UK, 2006; p. 125.

17. Makropoulos, C.K.; Butler, D. Distributed water infrastructure for sustainable communities. *Water Resour. Manag.***2010**, *24*, 2795–2816.

18. Sharma, A.; Cook, S.; Tjandraatmadja, G.; Gregory, A. Impediments and constraints in the uptake of water sensitive urban design measures in greenfield and infill developments. *Water Sci. Technol.* **2012**, *65*, 340–352.

19. Tjandraatmadja, G.; Burn, S.; McLaughlin, M.; Biswas, T. Rethinking urban water systems: Revisiting concepts in urban wastewater collection and treatment to ensure infrastructure sustainability. *Water Sci. Technol. Water Supply***2005**, *5*, 145–154.

20. Marleni, N.; Gray, S.; Sharma, A.; Burn, S.; Muttil, N. Impact of water source management practices in residential areas on sewer networks—A review. *Water Sci. Technol.* **2012**, *65*, 624–642.

21. Sapkota, M.; Arora, M.; Malano, H.; George, B.; Nawarathna, B.; Sharma, A.; Moglia, M. Development of a Framework to Evaluate the Hybrid Water Supply Systems. In Proceedings of the 20th International Congress on Modelling and Simulation, Adelaide, Australia, 1–6 December 2013; pp. 2387–2393.

22. Poustie, M.S.; Deletic, A.; Brown, R.R.; Wong, T.; de Haana, F.J.; Skinner, R. Sustainable urban water futures in developing countries: The centralised, decentralised or hybrid dilemma. *Urban Water* **2015**, *12*.

23. Cook, S.; Tjandraatmadja, G.; Marleni, N. *Impact of Source Management Strategies On quality and Loads in Residential Wastewater—Scenario Analysis*; CSIRO Water for a Healthy Country National Research Flagship: Melbourne, Australia, 2010.

24. Abrishamchi, A.; Ebrahimian, A.; Tajrishi, M.; Mariño, M.A. Case study: Application of multicriteria decision making to urban water supply. *J. Water Resour. Plan. Manag.* **2005**, *131*, 326–335.

25. Libralato, G.; Ghirardini, A.V.; Avezzù, F. To centralise or to decentralise: An overview of the most recent trends in wastewater treatment management. *J. Environ. Manag.* **2012**, *94*, 61–68.

26. Hering, J.G.; Waite, T.D.; Luthy, R.G.; Drewes, J.E.; Sedlak, D.L. A changing framework for urban water systems.*Environ. Sci. Technol.* **2013**, *47*, 10721–10726.

27. Coombes, P.J.; Cullen, A.; Bethke, K. Towards sustainable cities—Integrated water cycle management (iwcm) at the existing principal activity centre at doncaster hill. In Proceedings of the 34th World Congress of the International Association for Hydro-Environment Research and Engineering: 33rd Hydrology and Water Resources Symposium and 10th Conference on Hydraulics in Water Engineering, Brisbane, Australia, 26 June–1 July 2011; Valentine, E.M., Apelt, C.J., Ball, J., Chanson, H., Cox, R., Ettema, R., Kuczera, G., Lambert, M., Melville, B.W., Sargison, J.E., Eds.; Engineers Australia: Barton, Australia, 2011; pp. 2631–2638.

28. Graddon, A.R.; Kuczera, G.; Hardy, M.J. A flexible modelling environment for integrated urban water harvesting and re-use. *Water Sci. Technol.* **2011**, *63*, 2268–2278.

29. Graddon, A.R.; Kuczera, G.; Hardy, M.J. The modelling of urban water supply, harvesting and recycling systems using network linear programs. In Proceedings of the H2009: 32nd Hydrology and Water Resources Symposium, Barton, Australia, 30 November–3 December 2009; Engineers Australia: Newcastle, Australia, 2009; pp. 865–876.

30. Sharma, A.K.; Cook, S.; Chong, M.N. Monitoring and validation of decentralised water and wastewater systems for increased uptake. *Water Sci. Technol.* **2013**, *67*, 2576–2581.

31. Sharma, A.; Grant, A.L.; Grant, T.; Pamminger, F.; Opray, L. Environmental and economic assessment of urban water services for a greenfield development. *Environ. Eng. Sci.* **2009**, *26*, 921–934.

32. Cook, S.; Sharma, A.; Chong, M. Performance analysis of a communal residential rainwater system for potable supply: A case study in brisbane, australia. *Water Resour. Manag.* **2013**, *27*, 4865–4876.

33. Last, E.M. City Water Balance a New Scoping Tool for Integrated Urban Water Management Options. Ph.D. Thesis, The University of Birmingham, Birmingham, UK, 2010.

34. Mousseau, V. Eliciting information concerning the relative importance of criteria. In *Advances in Multicriteria Analysis*; Pardalos, P.M., Siskos, Y., Zopounidis, C., Eds.; Springer-Verlag: Berlin, Germany, 1995; pp. 17–43.

35. Mitchell, V.G.; Mein, R.G.; McMahon, T.A. Modelling the urban water cycle. *Environ. Model. Softw.* **2001**, *16*, 615–629.

36. Maheepala, S.; Leighton, B.; Mirza, F.; Rahilly, M.; Rahman, J. Hydro planner—A linked modelling system for water quantity and quality simulation of total water cycle. In *OzWater 07 Conference*; Australian Water Services Association: Sydney, Australia, 2007.

37. Stewardson, M.J.; Clark, R.D.S.; Cresswell, D.J.; McMahon, T.A. Modelling to assist integrated management of urban water resources. In Proceedings of the Second International Symposium on Urban Stormwater Management 1995: Integrated Management of Urban Environments, Melbourne, Australia, 11–13 July 1995; Institution of Engineers, Australia: Barton, Australia, 1995; pp. 107–112.

38. Hardy, M.J.; Kuczera, G.; Coombes, P.J. Integrated urban water cycle management: The urbancycle model. *Water Sci. Technol.* **2005**, *52*, 1–9.

39. Mitchell, V.G.; Diaper, C.; Gray, S.R.; Rahilly, M.; Technolgy, C.M.A.I. Uvq: Modelling the movement of water and contaminants through the total urban water cycle. In Proceedings of the 28th International

Hydrology and Water Resources Symposium, Wollongong, Australia, 10–13 November 2003; p. 8.

40. Clarke, L.; Lurz, J.; Wise, M.; Edmonds, J.; Kim, S.; Smith, S.; Pitcher, H. *Model Documentation for the Minicam Climate Change Science Program Stabilization Scenarios: Ccsp Product 2.1 A*; Pacific Northwest National Laboratory: Richland, WA, USA, 2007.

41. Snowdon, D.; Hardy, M.J.; Rahman, J.M. Urban developer: A model architecture for manageably building urban water cycle models spanning multiple scales. In Proceedings of the 19th International Congress on Modelling and Simulation, Perth, Australia, 12–16 December 2011.

42. Mitchell, V.G.; Diaper, C. Simulating the urban water and contaminant cycle. *Environ. Model. Softw.* **2006**, *21*, 129–134.

43. Lai, E.; Lundie, S.; Ashbolt, N.J. Review of multi-criteria decision aid for integrated sustainability assessment of urban water systems. *Urban Water J.* **2008**, *5*, 315–327.

44. Hajkowicz, S.; Collins, K. A review of multiple criteria analysis for water resource planning and management. *Water Resour. Manag.* **2007**, *21*, 1553–1566.

45. Mutikanga, H.E.; Sharma, S.K.; Vairavamoorthy, K. Multi-criteria decision analysis: A strategic planning tool for water loss management. *Water Resour. Manag.* **2011**, *25*, 3947–3969.

46. Moglia, M.; Kinsman, D.; Maheepala, S. *Multi-Criteria Decision Assessment Methods to Identify Total Water Cycle Management Strategies*; Urban Water Security Research Alliance: Queensland, Australia, 2012; Techincal Report No. 101.

47. Moglia, M.; Sharma, A.K.; Maheepala, S. Multi-criteria decision assessments using subjective logic: Methodology and the case of urban water strategies. *J. Hydrol.* **2012**, *452–453*, 180–189.

48. Pomerol, J.-C.; Barba-Romero, S. *Multicriterion Decision in Management: Principles and Practice*; Springer Science & Business Media: New York, NY, USA, 2000; Volume 25.

49. Brans, J.P.; Mareschal, B. The promcalc & gaia decision support system for multicriteria decision aid. *Decis. Support Syst.* **1994**, *12*, 297–310.

50. Brans, J.P.; Mareschal, B. Promethee methods. In *Multiple Criteria Decision Analysis: State of the Art Surveys*; Figueira, J., Greco, S., Eds.; Springer Science + Business Media, Inc.: New York, NY, USA, 2005; pp. 163–195.

51. Kodikara, P.N.; Perera, B.J.C.; Kularathna, M.D.U.P. Stakeholder preference elicitation and modelling in multi-criteria decision analysis—A case study on urban water supply. *Eur. J. Oper. Res.* **2010**, *206*, 209–220.

52. De Smet, Y.; Lidouh, K. An introduction to multicriteria decision aid: The promethee and gaia methods. In *Business Intelligence*; Aufaure, M.-A., Zimányi, E., Eds.; Springer: Berlin/Heidelberg, Germany, 2013; pp. 150–176.

53. Hayez, Q.; de Smet, Y.; Bonney, J. D-Sight: A new decision making software to address multi-criteria problems. *Int. J. Decis. Support Syst. Technol.* **2012**, *4*, 1–23.

54. Wong, T.H.F.; Brown, R.R. The water sensitive city: Principles for practice. *Water Sci. Technol.* **2009**, *60*, 673–682.

55. Xue, X.; Schoen, M.E.; Ma, X.C.; Hawkins, T.R.; Ashbolt, N.J.; Cashdollar, J.; Garland, J. Critical insights for a sustainability framework to address integrated community water services: Technical metrics and approaches. *Water Res.* **2015**, *77*, 155–169.

56. VicUrban. *Aurora Development Part 2*; Victoria Growth Area Authority: Whittlesea, Australia, 2007; p. 87.

57. Department of Transport, Planning and Local Infrastructure. *Victoria in Future 2014 (VIF 2014)*; Department of Transport, Planning and Local Infrastructure: Melbourne, Australia, 2014.

58. Sharma, A.; Grant, A.; Gray, S.; Mitchell, G. *Sustainability of Alternative Water and Sewerage Servicing Options—YVW; Kalkallo and Box Hill Pac Developments*; CSIRO Urban Water: Highett, Australia, 2005.

59. ABS. *Environmental Issues: Water Use and Conservation*; Australian Bureau of Statistics: Canberra, Australia, 2010.

60. Sharma, A.; Tjandraatmadja, G.; Grant, A.; Grant, T.; Pamminger, F. Sustainable sewerage servicing options for peri-urban areas with failing septic systems. *Water Sci. Technol.* **2010**, *62*, 570–585.

61. Metcalf and Eddy Inc.; Tchobanoglous, G.; Burton, F.L.; Stensel, H.D. *Wastewater Engineering: Treatment and Reuse*, 4th ed.; McGraw-Hill: New York, NY, USA, 2003.

62. FAO. *Wastewater Treatment and Use in Agriculture*; Food and Agriculture Organization of the United Nations: Rome, Italy, 1992.

63. Henze, M.; Comeau, Y. Wastewater characterization. In *Biological Wastewater Treatment: Principles, Modelling and Design*; Henze, M., van Loosdrecht, M.C.M., Ekama, G.A., Brdjanov, D., Eds.; IWA Publishing: London, UK, 2008; pp. 33–52.

64. Riahi, K.; Grübler, A.; Nakicenovic, N. Scenarios of long-term socio-economic and environmental development under climate stabilization. *Technol. Forecast. Soc. Chang.* **2007**, *74*, 887–935.

65. Meinshausen, M.; Smith, S.; Calvin, J.K.; Daniel, J.S.; Kainuma, M.L.T.; Lamarque, J.F.; Matsumoto, K.; Montzka, S.A.; Raper, S.C.B.; Riahi, K.; *et al.* The rcp greenhouse gas concentrations and their extensions from 1765 to 2300. *Clim. Chang.* **2011**, *109*, 213–241.

66. Mein, R.G.; Apostolidis, N. A simple hydrologic model for sewer inflow/infiltration. In Proceedings of the International Symposium on Urban Stormwater Management, Sydney, Australia, 4–7 February 1992; Barton, A., Ed.; Engineers Australia: Sydney, Australia; pp. 204–208.

67. Mankad, A.; Tapsuwan, S. Review of socio-economic drivers of community acceptance and adoption of decentralised water systems. *J. Environ. Manag.* **2011**, *92*, 380–391.

68. EPA Victoria. *Guidelines for Environmental Management: Use of Reclaimed Water*; EPA Victoria: Victoria, Australia, 2003. [

Chapter 8

A Graph-Theoretic Framework for Assessing the Resilience of Sectorised Water Distribution Networks

Manuel Herrera[1] ·Edo Abraham[2] ·Ivan Stoianov[2]

[1] EDEn - Department of Architecture and Civil Engineering, University of Bath, Bath, BA2 7AY, UK

[2] InfraSense Labs - Department of Civil and Environmental Engineering, Imperial College London, London, SW7 2AZ, UK

ABSTRACT

Water utilities face a challenge in maintaining a good quality of service under a wide range of operational management and failure conditions. Tools for assessing the resilience of water distribution networks are therefore essential for both operational and maintenance optimization. In this paper, a novel graph-theoretic approach for the assessment of resilience for large scale water distribution networks is presented. This is of great importance for the management of large scale water distribution systems, most models containing up to hundreds of thousands of pipes and nodes. The proposed framework is mainly based on quantifying the redundancy and capacity of all possible routes from demand nodes to their supply sources. This approach works well with large network sizes since it does not rely on precise hydraulic simulations, which require complex calibration processes and computation, while remaining meaningful from a physical and a topological point of view. The proposal is also tailored for the analysis of sectorised networks through a novel multiscale method for analysing connectivity, which is successfully tested in operational utility network models made of more than 100,000 nodes and 110,000 pipes.

KEYWORDS

Water distribution networks, Resilience Complex systems, Graph theory, Multiscale graph decomposition

1. INTRODUCTION

Resilience can be defined as the ability of a system to maintain and adapt its operational performance in the face of failures and other adverse conditions (Laprie 2005; Strigini 2012). Depending on the application domain, different approaches are applied to assess the resilience of man-made systems. There is no single widely used definition for quantifying the resilience of water distribution networks (WDNs), a common method is to formulate the hydraulic resilience as a measure of the ability of a network to maintain supply under failure conditions. Todini (2000) proposed a resilience index based on the steady state flow analysis of WDNs and dissipated energy; consequently, the resilience of a water network was defined using a measure of the available surplus energy. Recent work extends this approach; for example, Raad et al. (2010) modelled the outcomes of this index proposing to work with alternative approaches to the original in Todini (2000); this was done to overcome computational difficulties in optimising a network design problem using Todini's resilience index as an objective function. Therefore, by using parameters of physical significance in the supply as surrogates and not the real hydraulic details, di Nardo et al. (2013a) included these surrogate indices for optimizing the resilience of sectorised water distribution networks. Baños et al. (2011) proposed an extension to Todini (2000) by considering failures under various water demand scenarios. Other approaches based on a steady state hydraulic analysis include Prasad and Park (2004) who adapted Todini's index by incorporating the effects of both surplus pressure and reliability of supply loops assessed by the variability in the diameter of pipes connected to the same node. They also applied the method to a multi-objective problem for the optimal design and rehabilitation of a water distribution network. Jayaram and Srinivasan (2008) used the same method from Prasad and Park (2004) for the analysis of networks with multiple sources. An alternative approach was proposed by Wright et al. (2015) who assess the resilience of networks with dynamically configurable topologies by estimating what percentage of operational demand can be supplied when a disruptive event occurs.

A common constraint of the above approaches is that the combination of possible failure scenarios grows exponentially as the network becomes bigger (Berardi et al. 2014) together with possible inconsistencies and uncertainties associated with hydraulic simulations (Gupta and Bhave 1994). Trifunović (2012) explores hydraulic properties of the network based on the statistical analysis of common parameters under normal operation and proposed them as indices to assess the resilience of a water network. However, these statistical parameters may not be valid under failure conditions (Giustolisi et al. 2008). Another inherent drawback is the dependence on well calibrated (accurate) hydraulic models for large networks. Beyond the implicit non-linearity, the number of parameters involved in the hydraulic equations (Abraham and Stoianov2015) and their large number of possible combinations introduce high complexity to the WDN calibration problem (an NP-Hard problem, which cannot be solved in polynomial time Takahashi et al. (2010).) For large WDNs, these have an expensive computational cost since, even for moderate size networks, the number of possible failure scenarios grows exponentially and are approached using either some linearising assumptions or the use of heuristics for failure simulations (Berardi et al. 2014).

This paper assesses the resilience of WDNs from a topological perspective where properties such as network configuration and redundancy in connectivity are taken into account together with physical-based flow properties. This assessment is based on a computationally efficient implementation of the *K*-shortest paths algorithm (Eppstein 1998). It measures in a statistically robust way how every WDN demand node would be affected by disruptions in supply. This is done by analysing alternative paths between demand nodes and water sources (i.e. tanks and reservoirs). In addition, the paths are weighted by the hydraulic attributes of the supply routes to make the measures physically meaningful. Graph-theoretic indices have been applied to assess the resilience of water networks and used also as a surrogate measure in design problems (Yazdani et al. 2011). These surrogate indices are measures that correlate to physical/hydraulic based indices of resilience but are not necessarily based on physical indices whose parameters are either unavailable or are difficult to compute. These indices can be broadly classified into "statistical" and "spectral" measures (Gutiérrez-Pérez et al. 2013). Among the statistical indices, the meshedness coefficient (Buhl et al. 2006) provides an estimation of the topological redundancy of a network. Central-point dominance (Freeman 1977), gives a measurement of the network vulnerability to failures corresponding to central locations. Flow Entropy (Raad et al. 2010) measures the strength of supply to a node both in terms of the number of connections and their similarity. The so-called demand-adjusted entropic degree in Yazdani and Jeffrey (2012) is another alternative that uses demand on nodes and volume capacity on edges to compute a weighted entropic degree. More recently, (Liu et al. 2014) have extended the flow entropy measure to also consider the impact of pipe diameter on reliability.

The work also extends the proposed adaptation of *K*-shortest paths algorithm for the analysis of WDNs divided into sectors or District Metered Areas (DMAs) by utilizing multiscale network decomposition (Albert and Barabási 2002; Lee and Maggioni 2011). Demand nodes of a given DMA are aggregated into a sector-node, where a new graph (DMA-graph) is defined with vertices representing DMA areas and edges that abstract the sector-to-sector connectivity (Izquierdo et al. 2011). A mapping function is used to establish a relationship between the network graph of demand nodes and the DMA-graph, where every DMA-vertex (or sector-node) has new features such as number of customers, total demand, and an average pressure aggregated from information of individual nodes of the DMA. The physical connections between sectors, which consist of pipes and boundary valves, form the edges of this DMA-graph and define the level of its connectivity. Resilience assessments carried out on individual demand nodes in the DMAs are also linked to the corresponding sector-nodes, which are then used for a system-level resilience analysis. Considering both the network graph of demand nodes and a DMA-graph, a multiscale analysis of resilience is proposed.

2. A GRAPH THEORY PERSPECTIVE ON THE DEFINITION OF WDN RESILIENCE

The network connectivity of a WDN can be modelled as a nearly-planar mathematical graph1, $G = (V, E)$, where V (vertices) corresponds to n nodes and E (edges) corresponds to m pipes of the water system. This graph also has the particular characteristics that every node or edge should normally have at least one path of edges connecting it to a source node (tanks and/or reservoirs).

For a given WDN, our approach starts with computing a measure of resilience for each node by identifying the number of water sources a node is connected to and estimating how well that node is supplied by the available water sources. To accomplish this, all routes connecting a node *i* to all tanks and reservoirs are computed, and used to quantify the capacity or resistance of the corresponding routes. Since computing all routes which link nodes to sources is prohibitively expensive, our approach becomes feasible by restricting the computations to the *K* 'shortest' routes (Eppstein 1998) for the flow between a node *i* and its *j*-th sources $_j$ (*i*). In calculating this index, every route should be weighted by a surrogate measure of flow resistance. For example, the proposed surrogate flow resistance measure depends on the physical characteristics of connecting pipes as water travels from *j*-th source, $s_j(i)$, to *i* and it can be estimated by a measure proportional to the energy loss associated with a route of supply. The estimation of energy loss in water transportation is commonly done using the Darcy-Weisbach and Hazen-Williams formulae. Although both formulae depend on hydraulic measures like velocity and roughness, only the proportional dependence on topological constants of the network has been considered as no detailed hydraulic modelling is used. Energy loss across a link is proportional to $f \times L/D$; where *f* is the friction factor, *L* is the length of the pipe and *D* represents its diameter. This resistance measure is used to weight every path in the calculus of shortest paths, providing a surrogate measure for potential energy loss associated with each path connecting a node and its water source. The proposed resilience index for a node is

$$I_{GT}(i) = \sum_{s=1}^{S} \left(\frac{1}{K} \sum_{k=1}^{K} \frac{1}{r(k,s)} \right), \quad (1)$$

where *S* is the total number of sources in the network and *r*(*k*, *s*) is a surrogate measure of the energy loss (the resistance to water transport) associated with the kth path to source *s*. One such measure of energy loss could be approximated as:

$$r(k) = \sum_{m=1}^{M} f(m) \frac{L_m}{D_m}, \quad (2)$$

where M is the number of pipes in path k and $f(\cdot)$ estimates the friction factor by pipe age and material (Christensen 2009).

The weighted KSP algorithm returns infinity for Eq. 2 if there is no path of water distribution connecting the given node with a supply source. The contribution of such a source to the resilience index for a node in Eq. 1will be zero. Consequently, both the resistance of paths to sources and redundancy in supply are explicitly taken into account by the resilience index.

In Eq. 2, the flow resistance over K paths is averaged since using the shortest path only may not be sufficient to represent the connectivity and indirectly resilience. By averaging over the complete set of routes (i.e. $K \rightarrow 2^n$, where n is a very big number for operational networks (Eppstein 1998)), the true connectivity of a node to a source is measured. However, the computational complexity of this measure grows exponentially (O(2nn)O(2nn) for the proposed index) and it quickly becomes infeasible for medium to large networks. Averaging over the K-shortest paths is a computationally feasible approximation that takes into account a sufficiently large number of paths in a robust statistical method.

2.1 Validation of the Resilience Assessment Method

A previously published network C-Town (Ostfeld et al. 2012) is used as a validation case-study for the graph-theoretic method proposed. This is a medium size network that consists of 333 nodes, 429 pipes, 4 valves, 5 pump stations, 7 elevated tanks, and 1 reservoir. A graph-theoretical analysis of resilience in WDNs (Yazdani et al. 2011) identifies a node of special importance as those with high betweenness centrality index, which measures the relative number of shortest paths from all vertices to all others that pass through a node. Nodes with betweenness centrality values larger than 2 times the standard deviation (SD) from the network average computed over all nodes will be denominated as high betweenness nodes. This work considers "critical transfer nodes" to those of topological importance with respect to the flow distribution with a WDN. Thus, critical transfer nodes are, a priori, those belonging to the main trunk and those with high betweenness.

In the case of the proposed K-shortest path index, for a demand node i and source s, let

$$g(K;i,s) = \frac{1}{K}\sum_{k=1}^{K}\frac{1}{r(k)}, \tag{3}$$

represent the inverse of the resistance to supply in Eq. 1 averaged over K paths. Figure 1a shows the difference $g(K + 1) - g(K)$ for two nodes of varying centrality, A and B, where the latter belongs to the set of central nodes. In the two cases, the average of Eq. 3 taking K shortest paths converges when the number of paths is increased. That is, the difference on computing $g(K + 1)$ or $g(K)$ changes significantly as one moves away from using only the shortest path (i.e. $K = 1$) but not as much when one goes from K to $K + 1$ paths for larger values of K (over 30 paths in the case of Fig. 1a). Then, this measure is proposed as a decision criterion on the number of paths to be used in Eq. 1 to compute $I_{GT}(\cdot)$.

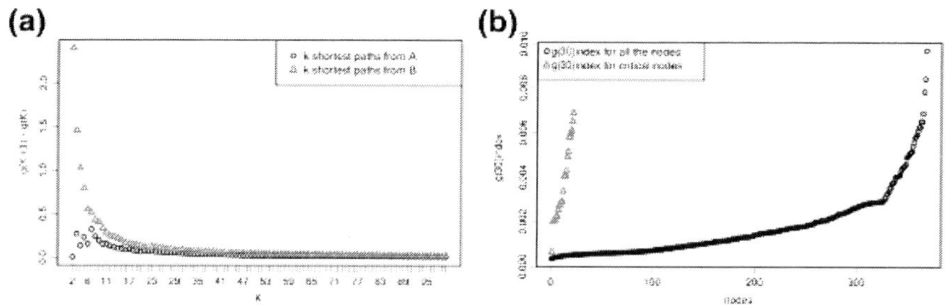

Figure 1. Validation analysis on the K routes to the water sources

Using K paths has two main advantages. Firstly, it emphasises the need to consider multiple paths of transport that contribute to the resilient operation of a network and it guarantees a robust statistical estimation since multiple paths are used to capture the average connectivity and capacity instead of the unique shortest path connecting two nodes. Secondly, the sum in Eq. 1 converges for relatively low values of K to obtain a good approximation of the analysis over all possible routes.

Once the tuning process provides an estimate of the K value, the proposal proceeds to compute the inverse of the resistance to supply, $g(K)$, for the chosen K. In Fig. 1b $g(30)$ is plotted for all individual nodes. The same plot also represents the values of $g(30)$ for central nodes (main-trunk nodes and those with a betweenness centrality of at least 2 times the SD of the network average; they are plotted with red triangles). Most of the critical nodes have high $g(K)$ compared to the average node $g(K)$ value. In general, nodes of higher betweenness centrality are those that will appear more frequently in any shortest path, including paths to sources, and so will tend to have higher resilience as defined by Eq. 1.

2.2 Comparison of K-Shortest Paths Index (I_{GT}) with Alternative Approaches

The next stage of the analysis is a comparison of the various indices of resilience under failure conditions in a supply network. Three scenarios generated from C-Town are considered, (Table 1, Fig. 1). The first scenario represents normal operational conditions without failures. The second scenario is created by the failure (removal) of 2.5 % of randomly selected pipes. The third scenario assumes the failure of 5 % of pipes. The resilience indices are expected to degrade as more pipes are taken off. The aim of this example is to evaluate the consistency of the new index, I_{GT}, by validating that it positively correlates with results obtained from hydraulic simulations. The comparison is done with Todini's hydraulic resilience index, I_r(Todini 2000). Other graph-theoretic indices, such are central-dominance betweenness (I_{CB}), meshedness (I_m), and a demand-adjusted entropic index (I_S) (Yazdani and Jeffrey 2012), are also included in this analysis. Table 1 summarises the average of the computed indices under the repetition of 20 random experiments for each scenario.

Table 1. Resilience indices computed when a percentage of pipes are removed

Index	0 % removed	2.5 % removed	5 % removed
I_r	0.4476	0.3697	0.2561
I_{GT}	0.6306	0.5711	0.4166
I_{CB}	0.0715	0.0620	0.0550
I_m	0.0597	0.0508	0.0419
I_S	0.0369	0.0355	0.0351

To compare the performance of the various indices, their relative change under failure conditions are investigated. The process starts with the computation of a nominal resilience value; these are the computed under normal operation of a WDN. Then, there are computed the indices under failure configurations and are plotted the ratio of the new resilience values to their corresponding nominal index. Figure 2a shows that the degradation in resilience is captured by the considered indices to a different degree. The simulation of failure indices in Fig. 2a is repeated 20 times (by randomly sampling the corresponding disrupted pipes at each repeated simulation; in this case, 20 samples is enough to arrange statistically meaningful comparisons between groups/indices). At each iteration, different sets of pipes (2.5 % and 5 %) are randomly sampled and the indices are averaged for each proportion of pipe failures in the WDN. Figure 2b shows the distribution of the indices in Fig. 2a under these multiple experiments.

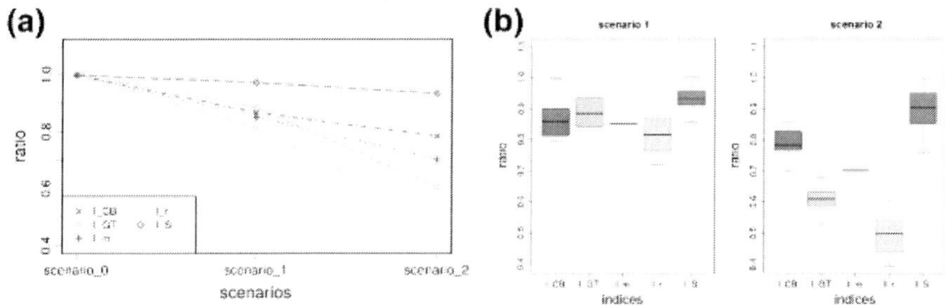

Figure 2. Graph-theoretic indices vs. hydraulic simulation based index

As shown in Fig. 2b, graph-theoretic indices show consistency with the hydraulic index2. The index I_{GT} changes in a proportional way to variations in the condition of the water distribution network. The entropic degree, I_S, is the measure which provides the poorest estimation of the deterioration in resilience as more pipes are removed, similarly to the results presented by (Greco et al. 2012). Figure 2b also shows that the meshedness index, I_m, has a narrow distribution. This is because meshedness is a measure that depends only on the relative number of nodes and pipes, without taking into account their physical or hydraulic properties.

3. ASSESSING RESILIENCE FOR LARGE SCALE WATER NETWORKS

The network topology of a WDN includes both tree-like (usually hierarchical and lower redundant networks, i.e. having an equivalent proportion of nodes and links connecting them) and well meshed structures, where there usually exists several paths connecting two nodes and so a greater number of links compared to the number of nodes. These two type of structures can be interconnected by links forming graph-bridges, which transport water from the transmission mains to these distribution areas. One way to analyse large-scale WDNs is by partitioning the network into different components (assets: valves, pumps, pipes, and meters, among others). Alternatively, a WDN can be divided into sub-networks in order to make the analysis simpler than working with the whole system model. This is of interest especially in the case of large-scale WDNs, which can contain hundreds of thousands of nodes and pipes, where consequently the classical methods of analysis are not efficient because of the large dimensions.

Many WDNs have been divided into sectors (DMAs) mainly for leakage management (Tabesh et al. 2009; Gomes et al. 2013; Xin et al. 2014). For this and a variety of other reasons (Herrera 2011), WDNs are divided into sectors nowadays and researches have focused their attention on different sectorisation approaches. The most studied works on water network division are based on variations of graph clustering (di Nardo and di Natale 2011; Perelman and Ostfeld 2011), spectral clustering (Herrera et al. 2010; Candelieri et al. 2014; Herrera et al. 2012), community detection (Diao et al. 2012), multi-agent systems (Izquierdo et al. 2011; Hajebi et al. 2013), breadth and depth first search (Deuerlein 2008; di Nardo et al. 2014) or multilevel partitioning (di Nardo et al. 2013b). The work of Mandel et al. (2015) is the first to propose a sectorisation approach for large-scale water networks. For these previously sectorised networks, the resilience index in Eq. 1 measures the level of connectivity of each demand node through various supply roots. A DMA division of the network automatically allows the use of two different scales of this resilience measure, having approaches of resilience for every individual node or measures summarised by sector. This provides new ways for analysing and managing the system and allows network configurations that have a major impact on resilience to be easily identified; for example multi-feed, single-feed, cascading DMAs.

3.1 Multiscale Resilience Analysis

The proposed graph-theoretic index is extended to water networks that are divided into sectors thus enabling resilience analysis. A DMA-based graph is a graph of lower level of detail (or coarser granularity) that captures the topology of existing DMAs. The DMA-scale resilience is derived from a user-defined function that transforms the resilience of the demand nodes in a DMA to a scalar representing the DMA resilience. For example, by transforming the index of Eq. 1 to the coarser sector level graph to analyse the relative connectivity of DMAs. Here, it is used the trimmed mean (García-Escudero et al. 2003) to transform the resilience of n∗ini∗ demand nodes of DMA_i to the sector graph as:

$$I^*_{GT}(DMA_i) = \sum_{j=1}^{n^*_i} \frac{I_{GT}(j)}{n^*_i}. \tag{4}$$

In calculating the mean, the trimmed approach discards points of very high or very low value before computing the mean for the remaining nodes. Then, the aggregated measure is the summary of the single values for n∗ini∗ nodes of the DMA, with n∗i≤nini∗≤ni. As such, a trimmed mean is a more robust centrality measure as the average value is not skewed by extreme low/high value data. For instance, if a DMA with predominantly high resilient nodes has a sub-area that has critically low resilience, the arithmetic average would not identify the presence of these in the DMA resilience index, whereas the trimmed method does. Furthermore, the presence of individual critical customers such as hospitals could significantly affect the system level resilience assessment. The set of excluded nodes in the computation of the trimmed mean, HH, are identified to validate the existence of critical nodes among them, and the subsequent impact on the DMA resilience assessment. As a result, Eq. 5 formulates a constitutive rule which is used to rank DMAs as low resilient if they have low values of I_{GT} for critical users (hospitals, schools, or industrial users) regardless of the fact that a DMA as a whole or a majority of its demand nodes might have high resilience.

$$I_{GT}(DMA_i) = \begin{cases} \min\{I_{GT}(H_i)\} & \text{if at least one critical user} \in \mathcal{H} \\ I^*_{GT}(DMA_i) & \text{otherwise,} \end{cases} \tag{5}$$

where H_i is the set of critical users in DMA_i.

For a DMA with a large variation in the I_{GT} value, it is possible to trim a significant number of nodes (i.e. a number of demand nodes representing a significant number of customers) in computing Eq. 4. To avoid this, the variability of the resilience is used, which is defined here as standard deviation normalized by the mean resilience value:

$$S^2_{GT}(DMA_i) = \sum_{j=1}^{n_i^*} \frac{1}{n_i^* - 1} \left(\frac{I_{GT}(j) - I^*_{GT}}{I^*_{GT}} \right)^2. \quad (6)$$

Based on the variability of $I_{GT}(\cdot)$ in each DMA_i (Eq. 6), a second constitutive rule is formulated and applied:

$$I_{GT}(DMA_i) = \begin{cases} \min_{DMA_i}\{I_{GT}(\cdot)\} & \text{if } S^2_{GT}(DMA_i) > t \\ I^*_{GT}(DMA_i) & \text{otherwise} \end{cases} \quad (7)$$

where t is a user-defined threshold.

When high variability exists, this constitutive rule computes the DMA resilience index over the whole set of demand nodes in the DMA and identifies the existence of low resilient nodes. The two Eqs. 5 and 7 assist operators to identify areas for a further investigation where a threshold of demand nodes or critical customers might not have the required level of resilience.

3.2 Multiscale Representation of a WDN

An essential part of the sector-level resilience analysis is the pre-processing of network data of varying types and complexity. For example, geographic information system (GIS) data, telemetry measurements, and hydraulic information is aggregated into an unified system that provides access to heterogeneous data sets. A novel augmented graph is proposed which transforms DMAs into sector-nodes. Each sector-node has aggregated data about the assets it contains, abstracting the high-granularity graph of hydraulic nodes to a low-granularity graph. The pipes and valves that connect different DMAs represent the edges between sector-nodes and inherit their corresponding properties. A schematic view of this multiscale decomposition process is represented in Fig. 3.

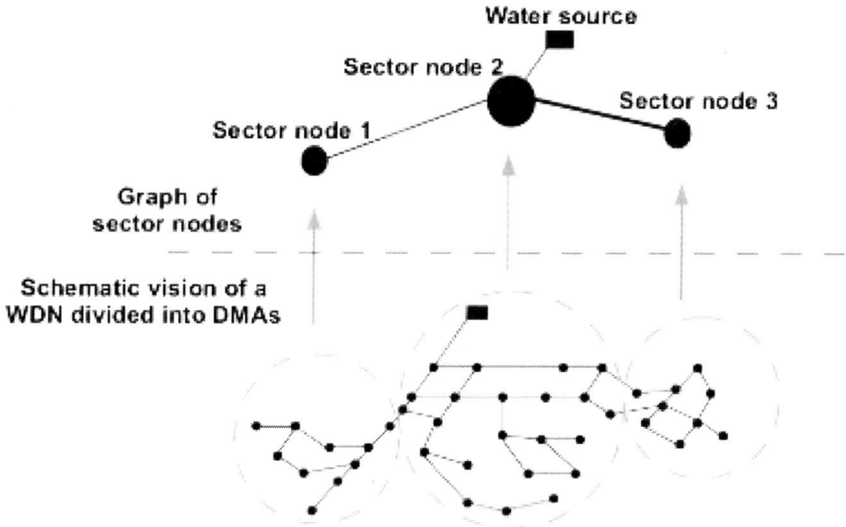

Figure 3. Mapping high-granularity demand-node into low-granularity DMA-based graphs

By considering all links hydraulically connecting DMAs, the edges of the new graph are formed by taking into account the open/closed status of valves, and consequently links which can provide additional operational resilience are identified. The weights of the new graph edges are proportional to the aggregated volumetric properties of individual pipes and valves that connect sector-nodes. Sector-nodes can also be used to visualize average levels of demand, pressure, elevation, etc. and the sector-node size can inherit topological characteristics of DMAs (eg. number of nodes or customers).

A multiscale network decomposition to assess the resilience of water systems is approached by studying their performance and network structure at various granularity levels and by analysing topological and physical information inherited from single demand nodes to sectors. Other important details that are obtained in the analysis include the number and location of multi-feed DMAs, the number of valves within each DMA, and the pressure differential between neighbouring DMAs. In addition, the sector-node graph is used to identify which part of the network topology should require further investigation to improve resilience.

4. EXPERIMENTAL STUDY

Two operational WDNs were used to analyse the proposed methodology for assessing the resilience of sectorised water distribution networks. Firstly, a medium size WDN divided into 3 DMAs is investigated. A resilience assessment is carried out for each DMA. The second case-study corresponds to a large water supply zone with high population density in England, UK. In both cases, certain sensitive information about the networks is withheld in order to maintain confidentiality.

4.1 Case-Study A

WDN-A (Fig. 4a) has 4820 nodes, 1 tank, and 5234 pipes and valves sectorised into 3 DMAs. Detailed information for each DMA is summarised in Table 2. Figure 4a also shows the DMAs layout and the location of critical transfer nodes (i.e. nodes with high values of central-betweenness).

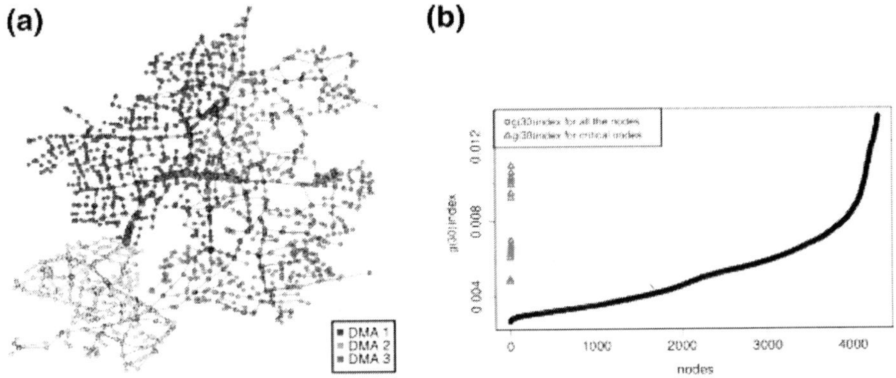

Figure 4. Layout and $g(K)$ index for the analysis of WDN-A

In order to compute the proposed graph-theoretic index, I_{GT} of Eq. 1, a sufficient number of shortest paths$K$ need to be used. The analysis outlined in Fig. 1 was carried out on different nodes and it showed that $K = 30$ as sufficient for WDN-A. In Fig. 4b, we plot the values of $g(30)$ for each node of the WDN and for demand nodes with high betweenness centrality (labelled as critical nodes and also marked in red in Fig. 4a). For this network, analysis in Fig. 4b shows that the measure $g(K)$ of the most critical nodes exceeds the average for the network.

Table 2. Summary of the DMA configuration and resilience indices for WDN-A

Sector ID	# nodes	# pipes	$I_{GT}(DMA)$
DMA-1	1626	1770	0.0654
DMA-2	2416	2638	0.0869
DMA-3	778	826	0.0465
TOTAL	4820	4032	0.0663

From Table 2, it can be concluded that DMA-3 is the sector with lowest resilience given the index $I_{GT}(DMA_3)$, while $I_{GT}(DMA_2)$ index is higher than the other two. DMA-2 is directly connected with the only source that supplies water to the whole network. Thus, topologically and hydraulically, DMA-2 is the most resilient sector of this network.

4.2 Case-Study B

WDN-B (shown in Fig. 5) is comprised of 106,115 nodes, 4 constant head sources, 28 reservoirs, 111,096 pipes, 102 pumps, and 1,900 boundary valves (Fig. 5a). GIS models, telemetry data, and physical and hydraulic information are combined in an integral database framework. Figure 5 shows the process of multiscale decomposition from the network of demand nodes to a coarser representation by sector-nodes. Through the graph transformation process, every DMA (sector-node) inherits the information related to the nodes within the area; such as, demand, elevation, pressure, and geographic coordinates. Sector-nodes also gather the information of the pipes and valve: diameter and length, and their own characteristics and the open/closed status for valves.

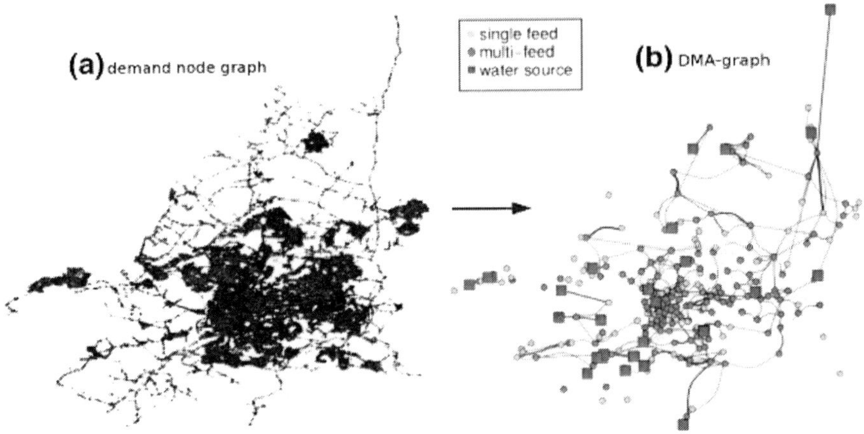

Figure 5. Layout of WDN-B and its representation by a graph of sector-nodes. Single-feed and multi-feed DMAs are differentiated visually.

The proposed multiscale decomposition of the network topology by sector-nodes reduces the dimensionality of complex networks. Figure 5b shows the spatial distribution of these sector-nodes together with their connectivity in terms pipes and valves. The volumetric capacity of aggregated links between DMAs are represented by the thickness of the connecting links. The colour code of the sector-nodes identifies whether a DMA is multi-feed (in colour red) or single feed (grey colour). The analysis shows that almost 40 % of the DMAs (i.e. 90 of the 228 DMAs) in WDN-B are fed by more than one supply connection, Fig. 5b. All this information together with the topology of the assets in the DMAs facilitate the estimation of the resilience indices at a DMA level, which is a new property of individual sector-nodes. Any measure based on individual assets within each DMA can be represented in the multiscale DMA-graph. Figure 6a shows the average resilience for each DMA, $I_{GT}(DMA_i)$, using a traffic light system. Red represents low resilience with an I_{GT} value 2 SDs below the network average for all DMAs; in this example, the network average is 0.0395 and 2 SDs below that is 0.02. Similarly, high resilience sectors (with values 2 SDs above average DMA resilience) are represented using green colour, and DMAs in between the two extremes with orange/amber colour. The augmented DMA-graph can also provide further information which assists operators in understating and analysing topological and operational constraints that affect resilience. Figure 6b highlights the critical transfer nodes of every DMA in distribution using the size of the circle plot associated with sector-nodes. In this figure, criticality is determined based on the number of high betweenness nodes.

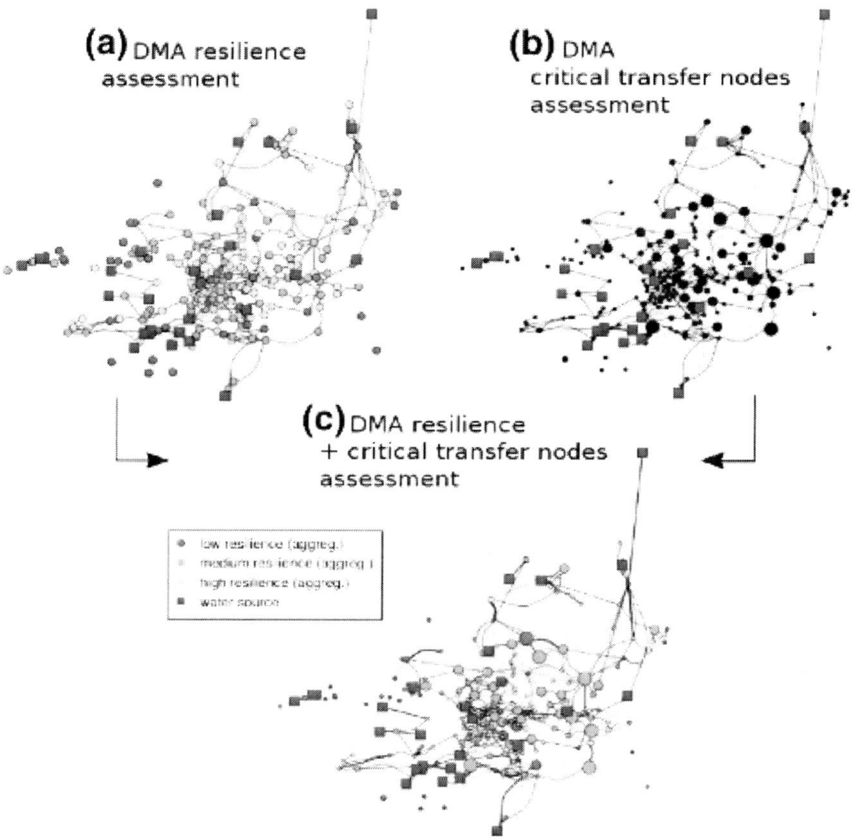

Figure 6. Schematic layout of WDN-B network resilience: $I_{GT}(DMA_i)$ given in Eq. (5) is represented by colour code (low=red, high=green) and the level of critical transfer by node size.

Resilience and critical transfer measures are simultaneously represented in Fig. 6(c), which combines a traffic light system and the sector-node size. As a result, the centre and centre-North of WDN-B include DMAs that have medium and low resilience measures and medium and high levels of flow transferability for central nodes (see Table 3). All these areas also supply critical customers and could be prioritised for further analysis and maintenance to improve network resilience. The boundaries of the WDN layout have low resilience but also low level of criticality. Table 3 presents a summary of attributes for the top 5 sectors in terms of their resilience and flow transfer critical nodes.

Table 3. Summary of the DMA configuration of the top 5 WDN-B sectors

Sector ID	# nodes	# pipes	Multi-feed	$I_{GT}(DMA)$	#CNs
DMA-29	2003	2081	Yes	0.0124	118
DMA-184	1119	1242	No	0.0179	32
DMA-85	784	834	Yes	0.0194	25
DMA-101	2106	2181	No	0.0212	60
DMA-208	2081	2249	Yes	0.0268	114

The resilience index, I_{GT}, is computed for each DMA. The number of critical transfer nodes (i.e. the number of nodes with betweenness centrality values larger than 2 times the SD from the average computed over all the DMA nodes) is given by #CNs

As shown in Table 3, where the DMAs are ranked by resilience, 3 of the 5 DMAs contain more than 2,000 nodes, while only one of them has a size less than 1,000 nodes. Taking into account the number of critical transfer nodes within each sector (column #CNs of Table 3), it is shown that DMA-29, DMA-101, and DMA-208 are sectors with larger number of transfer nodes. These DMAs also have a varying level of resilience; for example DMA-29 has the smallest resilience index and DMA-208 has the highest.

5. CONCLUSIONS

A novel resilience index based on weighted *K*-level shortest paths from consumption nodes to water sources has been derived to estimate WDN resilience together with information about critical customers. Computing these *K* different routes for every node provides numerical robustness to our analysis compared to analysing only the shortest path. The index is shown to be consistent with hydraulically-defined resilience measures. This is because the *K*-shortest paths are weighted by quantities associated with energy losses of water transport routes. The current proposal has the following advantages:

- An inclusion of physical attributes of the water distribution network within graph theoretic indices gives the resilience analysis a hydraulically relevant and easy to quantify measure of how well connected a node is to the available water sources.

- Computations for the proposed index scale quasi-linearly as a function of the number of nodes ($\sim O(m+n\log(n)+kn) \sim O(m+n\log(n)+kn)$). This makes the measure feasible for application to large scale water networks, where the classical critical link hydraulic based analysis of network resilience would scale combinatorially.

- Constitutive rules and functions facilitate the inclusion of empirical knowledge.

The proposed graph-theoretic framework is complimented by a novel multiscale decomposition method that converts the original WDN layout to DMA-based graphs. The developed method allows us to work with a WDN at two different levels of abstraction: single demand node and DMA. This enhances the results obtained for assessing the resilience of a WDN, especially in the case of large scale networks.

FOOTNOTES

1 A planar graph is one that can be drawn on a plane without having edges intersecting other than at a node mutually incident with them. WDN graphs are not strictly planar due to the existence of some possible crossovers but are almost planar.

2 Please note that the proposed index does not replace/estimate hydraulically simulated indices. The presented analysis validates the consistency of results as failure conditions change.

ACKNOWLEDGMENTS

This work has been performed with the support of the NEC - Imperial research project on Big Data Technologies for Smart Water Networks.

REFERENCES

1. Abraham E, Stoianov I (2015) Sparse null space algorithms for hydraulic analysis of large scale water supply networks. J Hydraul Eng:04015058

2. Albert R, Barabási AL (2002) Statistical mechanics of complex networks. Rev Mod Phys 74(1):47

3. Baños R, Reca J, Martínez J, Gil C, Márquez AL (2011) Resilience indexes for water distribution network design: a performance analysis under demand uncertainty. Water Resour Manag 25(10):2351–2366

4. Berardi L, Ugarelli R, Røstum J, Giustolisi O (2014) Assessing mechanical vulnerability in water distribution networks under multiple failures. Water Resour Res 50(3):2586–2599

5. Buhl J, Gautrais J, Reeves N, Solé RV, Valverde S, Kuntz P, Theraulaz G (2006) Topological patterns in street networks of self-organized urban settlements. The European Physical Journal B - Condensed Matter and Complex Systems 49(4):513–522

6. Candelieri A, Conti D, Archetti F (2014) Improving analytics in urban water management: a spectral clustering-based approach for leakage localization. Procedia-Social and Behavioral Sciences 108:235–248

7. Christensen RT (2009) Age effects on iron-based pipes in water distribution systems. Utha State University, PhD thesis

8. Deuerlein JW (2008) Decomposition model of a general water supply network graph. J Hydraul Eng 134(6):822–832

9. di Nardo A, di Natale M (2011) A heuristic design support methodology based on graph theory for district metering of water supply networks. Eng Optim 43(2):193–211

10. di Nardo A, di Natale M, Santonastaso GF, Tzatchkov VG, Alcocer-Yamanaka VH (2013a) Water network sectorization based on a genetic algorithm and minimum dissipated power paths. Water Science & Technology: Water Supply 13(4):951–957

11. di Nardo A, di Natale M, Santonastaso GF, Venticinque S (2013b) An automated tool for smart water network partitioning. Water Resour Manag 27(13):4493–4508

12. di Nardo A, di Natale M, Santonastaso G, Tzatchkov V, Alcocer-Yamanaka V (2014) Water network sectorization based on graph theory and energy performance indices. J Water Resour Plan Manag 140(5):620–629

13. Diao K, Zhou Y, Rauch W (2012) Automated creation of district metered area boundaries in water distribution systems. J Water Resour Plan Manag 139(2):184–190

14. Eppstein D (1998) Finding the k shortest paths. SIAM J Comput 28(2):652–673

15. Freeman LC (1977) A set of measures of centrality based on betweenness. Sociometry 40(1):35–41

16. García-Escudero LA, Gordaliza A, Matrán C (2003) Trimming tools in exploratory data analysis. J Comput Graph Stat 12(2)

17. Giustolisi O, Kapelan Z, Savic D (2008) Algorithm for automatic detection of topological changes in water distribution networks. J Hydraul Eng 134(4):435–446

18. Gomes R, Marques AS, Sousa J (2013) District metered areas design under different decision makers options: Cost analysis. Water Resour Manag 27(13):4527–4543

19. Greco R, di Nardo A, Santonastaso G (2012) Resilience and entropy as indices of robustness of water distribution networks. J Hydroinformatics 14(3):761–771

20. Gupta R, Bhave PR (1994) Reliability analysis of water-distribution systems. J Environ Eng 120(2):447–461

21. Gutiérrez-Pérez J, Herrera M, Pérez-García R, Ramos-Martínez E (2013) Application of graph-spectral methods in the vulnerability assessment of water supply networks. Math Comput Model 57(7–8):1853–1859

22. Hajebi S, Barrett S, Clarke A, Clarke S (2013) Multi-agent simulation to support water distribution network partitioning. In: 27th european simulation and modelling conference - ESM'2013. Lancaster University, UK, pp 1–6

23. Herrera M (2011) Improving water network management by efficient division into supply clusters. PhD thesis, Hydraulic Eng. and Environmental Studies, Universitat Politecnica de Valencia

24. Herrera M, Canu S, Karatzoglou A, Pérez-García R, Izquierdo J (2010). In: Swayne DA, Yang W, Voinov A, Rizzoli A, Filatova T (eds) An approach to water supply clusters by semi-supervised learning

25. Herrera M, Izquierdo J, Pérez-García R, Montalvo I (2012) Multi-agent adaptive boosting on semi-supervised water supply clusters. Adv Eng Softw 50:131–136

26. Izquierdo J, Herrera M, Montalvo I, Pérez-García R, Cordeiro J, Ranchordas A, Shishkov B (2011) Division of Water Supply Systems into District Metered Areas Using a Multi-agent Based Approach. Software and data technologies, communications in computer and information science, vol 50, Springer Berlin Heidelberg, pp 167–180

27. Jayaram N, Srinivasan K (2008) Performance-based optimal design and rehabilitation of water distribution networks using life cycle costing. Water Resour Res 44(1):1–15

28. (2005) Resilience for the scalability of dependability. In: Laprie JC (ed) Fourth IEEE International Symposium on Network Computing and Applications, pp 5–6

29. Lee JD, Maggioni M (2011) Multiscale analysis of time series of graphs. In: International Conference on Sampling Theory and Applications (SampTA)

30. Liu H, Savić D, Kapelan Z, Zhao M, Yuan Y, Zhao H (2014) A diameter-sensitive flow entropy method for reliability consideration in water distribution system design. Water Resour Res 50(7):5597–5610

31. Mandel P, Maurel M, Chenu D (2015) Better understanding of water quality evolution in water distribution networks using data clustering. Water Res 87(15):69–78

32. Ostfeld A, Salomons E, Ormsbee L, Uber J, Bros C, Kalungi P, Burd R, Zazula-Coetzee B, Belrain T, Kang D, Lansey K, Shen H, McBean E, Yi Wu Z, Walski T, Alvisi S, Franchini M, Johnson J, Ghimire S, Barkdoll B, Koppel T, Vassiljev A, Kim J, Chung G, Yoo D, Diao K, Zhou Y, Li J, Liu Z, Chang K, Gao J, Qu S, Yuan Y, Prasad T, Laucelli D, Vamvakeridou Lyroudia L, Kapelan Z, Savic D, Berardi L, Barbaro G, Giustolisi O, Asadzadeh M, Tolson B, McKillop R (2012) Battle of the water calibration networks. J Water Resour Plan Manag 138(5):523–532

33. Perelman L, Ostfeld A (2011) Topological clustering for water distribution systems analysis. Environ. Model Softw 26(7):969–972

34. Prasad T, Park N (2004) Multiobjective genetic algorithms for design of water distribution networks. J Water Resour Plan Manag 130(1):73–82

35. Raad DN, Sinske AN, van Vuuren JH (2010) Comparison of four reliability surrogate measures for water distribution systems design. Water Resour Res 46(5):1–11

36. Strigini L (2012) Fault tolerance and resilience: meanings, measures and assessment. In: Wolter K, Avritzer A, Vieira M, van Moorsel A (eds) Resilience Assessment and Evaluation of Computing Systems. Springer, Berlin, pp 3–24

37. Tabesh M, Yekta AA, Burrows R (2009) An integrated model to evaluate losses in water distribution systems. Water Resour Manag 23(3):477–492

38. Takahashi S, Saldarriaga J, Vega M, Hernández F (2010) Water distribution system model calibration under uncertainty environments. Water Science & Technology: Water Supply

39. Todini E (2000) Looped water distribution networks design using a resilience index based heuristic approach. Urban Water 2(2):115–122

40. Trifunović N (2012) Pattern recognition for reliability assessment of water distribution networks. PhD thesis, UNESCO-IHE Institute for Water Education, Delft University of Technology

41. Wright R, Abraham E, Parpas P, Stoianov I (2015) Control of water distribution networks with dynamic DMA topology using strictly feasible sequential convex programming. Water Resources Research 51(12):9925–9941.

42. Xin K, Tao T, Lu Y, Xiong X, Li F (2014) Apparent losses analysis in district metered areas of water distribution systems. Water Resour Manag 28(3):683–696

43. Yazdani A, Jeffrey P (2012) Water distribution system vulnerability analysis using weighted and directed network models. Water Resour Res 48(6):W06,517

44. Yazdani A, Otoo RA, Jeffrey P (2011) Resilience enhancing expansion strategies for water distribution systems: a network theory approach. Environ Model Softw 26(12):1574–1582

Chapter 9

Optimization of the Designed Water Distribution System Using MATLAB

Al-Amin D. Bello[1], Waheed A. Alayande[2], Johnson A. Otun[3], Abubakar Ismail[3], Umar F. Lawan[1]

[1]Faculty of Civil Engineering, Universiti Teknologi Malaysia, Johor Bahru, Malaysia

[2]National Water Resources Institute (NWRI) Mando, Kaduna State, Nigeria

[3]Department of Water Resources and Environmental Engineering, Ahmadu Bello University Zaria, Nigeria

ABSTRACT

This paper provides a technique of cost optimization for the proposed water distribution system before implementation. A lot of existing water distribution network analysis software lack optimization modules but ensure other essential conditions are satisfied. Dukku Town chose as a case study, it suffers from poor water distribution network which called for the system upgrade. The models developed by Alperovits and Shamir (1977) and modified by Goulter and Coal (1986) were used to accomplish the desire objectives. A programme is written in MATLAB using linear optimization components to compute the least cost possible. Changes in pipe diameters were obtained and there is slight significant decrease in total cost of pipes. A total of 7.15% reductions in the initial cost are noted, as the hydraulic properties of the entire distribution network is improved.

The maximum pressure before optimization is 14.79m and after optimization increased to 15.71m, the minimum pressure on the former is 7.59m and 9.42m for the later. The same improvement is observed in the flow rate, velocities and Headloss, all falls within the designed criterion. The efficiency of the network and risk is also considered by incorporating reliability constraint. The final optimized designed water distribution network addresses the problem of the water shortage of the studied area.

KEYWORDS

Optimization, Linear programming, MATLAB, Statistical Analysis, Dukku town

1. INTRODUCTION

A water distribution system (WDS) is a hydraulic conveyance system laid on road shoulders where topology and topography are known and that transmit water from the Source to the consumers; it consists of elements such as pipes, valves, pumps, tanks and reservoirs, flow regulating and control devices [1]. Water supply system efficiency is of primary importance in designing either new water distribution networks or expanding existing ones [2]. A WDS is normally designed and operated to satisfy various customer demands over its service life. Decision makers always explore innovative and efficient strategies to reduce the huge economic requirements of designing and construction of WDS, coupled with satisfying the quantity and performance objective of the system [3]. With significant development of urbanized areas and construction of thousands of small and large-scale water supply and distribution systems in recent decades, yet few people have access to clean water and adequate sanitation. However the quality of service provided by water utilities are unsatisfactorily and the cost of new systems are expensive [4]. In order to provide a reliable framework for system's operation, it is necessary to identify the most critical system requirement with a least cost that will enhance proper deliverance and effective operational management [5]. A lot of Research had been conducted to achieve these objectives [6-12]. Although pipe sizes and cost are the most important components of a water supply networks but other element are also considered because any water distribution network consists of three major parts and components:

namely pumps, distribution storages, and distribution piping network [13]. Although in optimizing a system, the designer must take some expected and unexpected loading conditions into consideration to ensure effective and sufficient deliverance of water to end user [14].

Selection of pipe diameters from a set of commercially available diameters to form a water distribution network of least capital cost has been shown to be a difficult task [1]. The cost of maintenance and operation of a water distribution system may be considerable, but still one of the main costs is that of the pipelines. The use of optimization methods for WDS has been mostly discussed over the last decades. Optimizing WDS involves the resolution of two issues: such as design and operations [14]. Design optimization usually deals with pipes sizes while WDS operation optimization takes into account of the pump schedules [15].

In recent years a number of optimization techniques have been developed mainly for the cost minimization aspect of network planning, although some reliability studies and stochastic modelling of demands have been attempted [1]. Some of the earlier studies utilized linear programming [16-18], while later studies applied nonlinear programming (NP) [19-21] or chance constrained [20] to the pipe network optimization problem. Much of the recent literature has utilized genetic algorithms for the determination of low cost water distribution network design and they have been shown to have several advantages over more traditional optimization methods [22, 23]. Linear optimization methods have been widely studied for the case of determining optimal design of water distribution networks [24, 25].

2. MATERIAL AND METHODS

2.1. Description of the Study Area

Dukku is the headquarter of Dukku Local Government Area (LGA) of Gombe State (Fig. 1). The LGA has an Area of 3,815km^2 and Population of 207,190 as at 2006 census. Dukku is located at Latitude 10.82° N and Longitude 10.77° E with Elevation of 608 m above sea level.

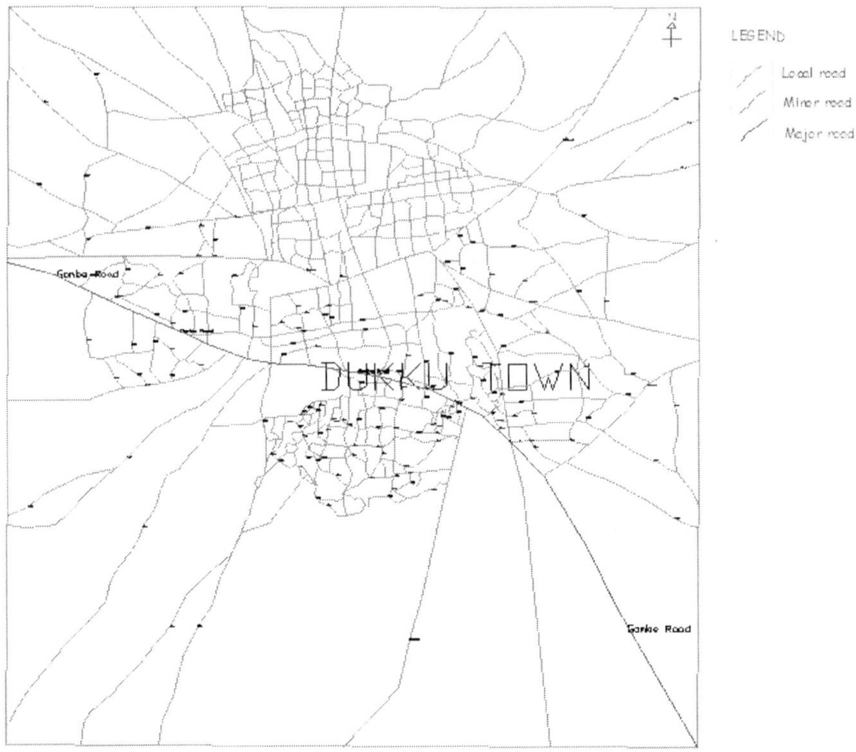

Figure 1. Dukku town

2.2. Dukku Water Supply System

The Dukku water supply source is purely underground water source and the town draws its water supply from a network of five boreholes drilled on the floodplains of the River Abba along the Gombe – Darazo road. The first three sets of borehole were drilled in 1975 and two additional one were drilled by the Upper Benue River Basin Development Authority. Average water discharge from the boreholes is about 320m^3/hr.

The boreholes discharge water through a 15km long 300mm diameter ductile iron rising mains to the booster station near *Wuro-Tara*. The booster pumps also pump through another 15km rising mains 250mm ductile iron pipe to deliver water to Dukku. The initial design of WDS layout before optimization is shown in figure 2.

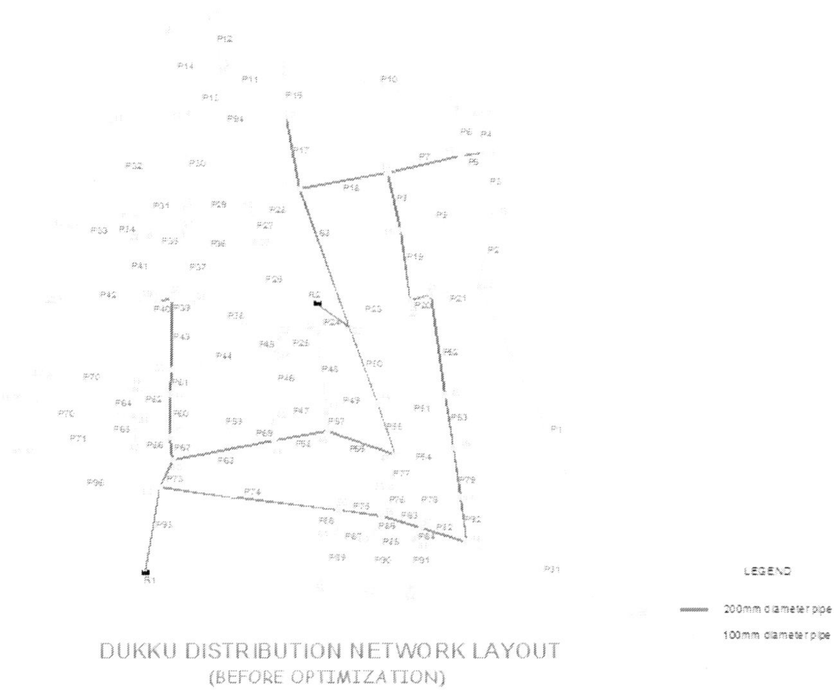

Figure 2. WDN Layout before optimization [26]

2.3. Software and Programming Tool Used

Water Distribution Network optimization analysis have broad components and complexity in terms of analysis and problem solving [27]. Therefore, a lot of programmes are developed to solved those problems and it required flexible, simple to write and easier to understand concepts. For this study MATLAB (matrix laboratory) developed by MathWork [28] was_used because it can solve many technical computational problems, especially those with matrix and vector formulations, in a fraction of the time it would take to write a program in a scalar non-interactive language as compared to C or FORTRAN [29]. EPANET water distribution network analysis software is employed to analyze the network. This solver uses the basic hydraulic principles to solve and analyze WDN [30], of which the results obtained in the solver are prepared as an input file for MATLAB to Optimized.

2.4. Optimization Procedure

This approach, which seeks to determine the pipe sizes and associated lengths so as to minimize the cost of the system while satisfying hydraulic criteria and reliability requirements, is derived from a model developed by Goulter and Coals [31] which in turn is originated from an earlier model developed by Alperovits and Shamir [16] with slight modification, as shown below.

Objective Function:

Minimization of network Cost

The objective function used to minimize the network cost for the Dukku water supply system formulated as [16, 29] given in equation 1

$$\text{Cost}(c) = \sum_{j=1}^{Nl} \sum_{k=1}^{n(j)} C_{jk} X_{jk} \tag{1}$$

Where; C : total cost of the system (N)

C_{jk} : cost of pipe of diameter k in link j (N/km)

X_{jk} : length of pipe of diameter k in link j (km)

N_L : total number of links within the system

$n_{(j)}$: number of different pipe diameters in link j

j : link index

k : diameter type index

Subject to the following constraints:

i. Length: the sum of the lengths of pipe in each link must equal the total length of the link where a link represents a pipe connecting two nodes directly as shown in equation 2.

$$\sum_{k=1}^{nj} X_{jk} = L_{oj} \tag{2}$$

For all link j,

Where; L_{oj} : total length of link j (km)

ii. Head loss: minimum and maximum permissible head at each demand point or Node must be satisfied (i.e Minimal and Maximal Pressures) shown in equation 3 and 4

$$H_o - \sum_{j=p(n)} \sum_{k=1}^{n(j)} J_{jk} . X_{jk} \geq H_{nmin} \tag{3}$$

$$H_o - \sum_{j \Xi p(n)} \sum_{k=1}^{n(j)} J_{jk} . X_{jk} \leq H_{nmax} \tag{4}$$

For all nodes n

Where; H_o : original head at source (m)

J_{jk} : hydraulic gradient for pipe diameter k in link j (m/km)

H_{nmin} : minimum allowable head at node n (m)

H_{nmax} : maximum allowable head at node n (m)

$p_{(n)}$: links in the path from source to node n

iii. Head loss: the total head loss must be equal the sum of head losses of all the pipes in series shown in equation 5.

$$\sum_{j \epsilon p(b)} \sum_{k=1}^{n(j)} J_{jk} . X_{jk} = J_{oj} . L_{oj} \tag{5}$$

Where; J_{oj} : Total Headloss within the link j (m/km)

iv. Non-negativity: Is assume all the pipe length are positive as shown in equation 6

$$X_{jk} \geq 0 \tag{6}$$

for all j and k

v. Reliability: an estimate of reliability is included into this constraint put by equation 7 which limit the expected (average) number of leakages or breaks in given time period in any link within the network.

$$\sum_{k=1}^{nj} R_{jk} \cdot X_{jk} \prec R_{oj} \tag{7}$$

Where;

R_{jk}: expected number of breaks/km/year for diameter k in link j

R_{oj}: maximum allowable number of failures per year in link j

The steps of the solution are summarized in the following points:

Step 1 of the algorithm: preparing the input file for EPANET

The designed water distribution network of Dukku were extracted and prepared as input file for EPANET and saved as Dukku.inp. The text input file contains network information such as:

i. Junction elevations and flow demands,

ii. Reservoir elevations,

iii. Pipe lengths, assumed diameters, roughness, and topology properties.

iv. Hydraulic analysis options.

Input.inp file was created from EPANET command menu (Fig 3).

Step 2 of algorithm: Network hydraulic analysis by EPANET

In this step, EPANET analyzes network defined by Dukku.inp file.

Step 3 of algorithm: Output of hydraulic analysis by EPANET

In this step output of the hydraulic analysis is obtained and saved.

Step 4 and step 5 of algorithm: Prepare input file for LOP – MATLAB

The Linear Optimization Programming (LOP) required the basic input from the solver which covered the flow rate for each pipe in between two nodes, the generated Unit head loss per link and the length of each link. The LOP uses the Hazen William equation to determine the unit head loss for the alternative diameters which is in consonant with the EPANET solver which uses the same Equation in the analysis.

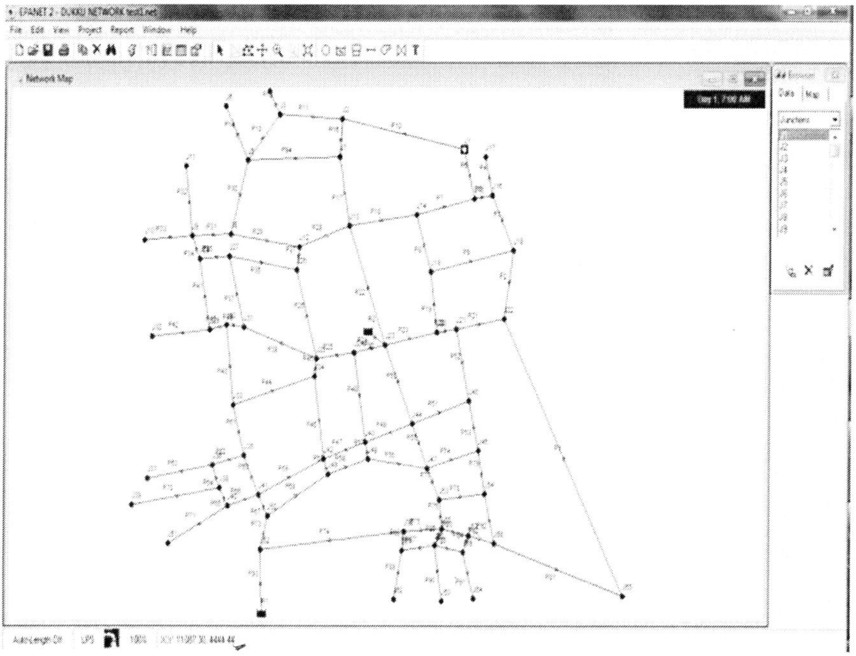

Figure 3. Showing Dukku Network in EPANET Interface

Objective function and constraint equations try to find best diameters for network links to reach the optimum result. Note that, in this approach, assumed unknown parameter for any link is not the pipe diameter but the lengths of the available pipe diameters.

In the study its assume that predefined available diameters for any sample link is: 100, 150, 200 and 250 mm; and associated cost and lengths for these diameter types are C_1, C_2, C_3, C_4 and X_1, X_2, X_3,X_4

Objective function for any sample link is:

$F = C_1*X_1 + C_2*X_2 + C_3*X_3 + C_4*X_4$

Linear Optimization Programme (LOP) defines it as matrix of *mxn* of which C_1, C_2, C_3, C_4, C_5,C_6 have a define value (CONSTANT) and X_1, X_2, X_3, X_4, X_5, X_6 are the Variables subject to the Constraints

Constraints 1: LOP defines it as Equality Constraints of mxn matrix, in reference to Equation 2.0.

$X_1 + X_2 + X_3 + X_4 = L$

Where L is the Total length per Link*Constraints 2:* LOP defines it as a Matrix of define Rows and Column and it's called Inequality Constraints, see the Equations (3.0) and (4.0), reproduced respectively as

For easier conversion to matrix form its changes to;

$$X_1h_1+X_2h_2+X_3h_3+X_4h_4 \leq (H_0-H_{min}) - (X_1h_1+X_2h_2+X_3h_3+X_4h_4) \leq (H_{max}-H_0)$$

And, H_o-H_{min} = H, H_{max}-H_o=H_L

Where;

H_0 = Head of the Reservoir (m)

H_{max} = Maximum allowable head (m) in the Network-User define

H_{min} = Minimum allowable head (m) in the Network-User define

H_1, h_2, h_3, h_4 = Are the respective unit head loss of the sample diameter of corresponding lengths X_1, X_2, X_3, X_4 (Determine by LOP using Hazen William Equation) in the Programme the *H* and H_L are User-define

Constraints 3: It is also called Equality constraint in LOP programme, as shown in Equation (5.0)

Is converted as; $X_1h_1+X_2h_2+X_3h_3+X_4h_4 = L_oH$

Where;

L = total length of the link in view

H = Head loss generated by the EPANET corresponding to the link in view

Constraints 4: LOP defines it as Bound constraints, Equation (6.0). This constraint makes all the variables in the forms

$$X_1 \geq 0 \qquad X_2 \geq 0 \qquad X_3 \geq 0 \qquad X_4 \geq 0$$

Constraints 5: LOP defines it as a Matrix of define Rows and Column and its called Inequality Constraints, Equation (7.0)

It is converted to matrix form in this format

$$X_1r_1+X_2r_2+X_3r_3+X_4r_4 \leq R$$

Where;

R = it is the expected number of break (greater than or equal to zero) per year in the whole network (user define)

r_1,r_2,r_3,r_4 = the expected number of breaks per km per year corresponding to the variables X_1,X_2,X_3,X_4.

The program is run, and the output result is display in a text format

Step 6, 7, 8 and 9 of algorithm: output the result obtained in MATLAB code

After the result is obtained in the text format, the respective diameters of each link is observed and the diameter with higher length is chosen or if the length is relatively close to the higher length but shifted significantly on the diameter that is higher than the initial diameter, the observed changes are then updated and the solver is re-run again. The result obtained in the solver is compare to the previous result and if all condition are satisfied then optimization done. The Condition includes

i. Are the design condition satisfied

ii. Is the flow direction constant

These procedures are maintained for all the links and updated in the Solver until all the conditions are satisfied.

3. RESULT AND DISCUSSION

The result shows additional 150mm and 250mm size pipes in the network. The network in the end has four different sizes of pipes (Water Distribution Network layout of the study area is updated as shown in fig. 3) unlike the initial designed that considered two different sizes, 100mm and 200mm only (fig. 2).

In other to ascertain the significant of the optimization, a statistical analysis was used and the results obtained are explained according to the desired parameters and importance.

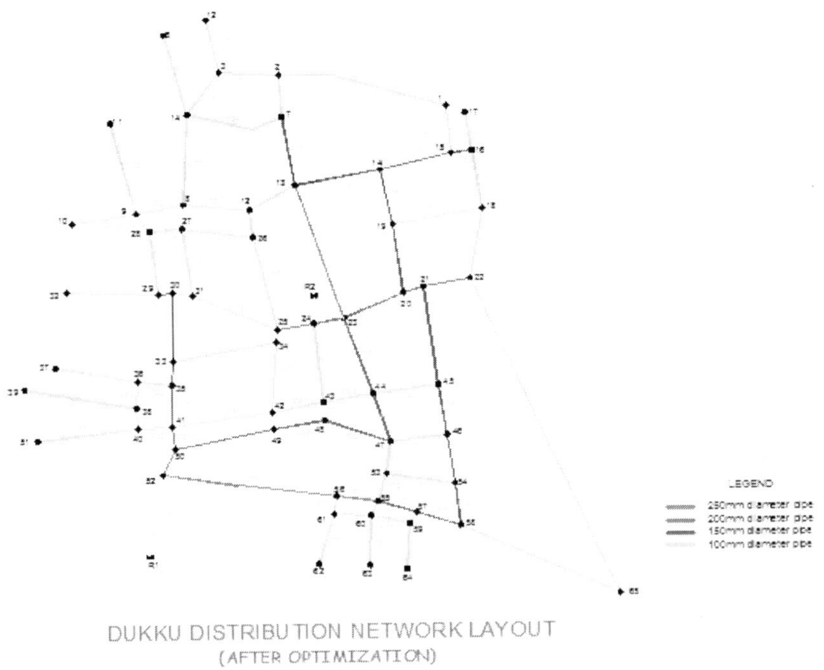

Figure 4. WDN layout after Optimization

Cost: A statistical comparison of the cost of pipes before and after optimization using non-parametric analysis (The Populations are not normally distributed) based on Wilcoxon test type and Monte Carlo Confidence interval of 95% was computed, and the result obtained is shown in Tables 1.0 and 2.0. The null hypothesis is that, there is no difference between the cost of pipes of our sample prior to optimization of the Network and the final cost obtained after optimization of the Network. Using two related sample method, the "Positive Ranks" are have a much greater sum than the negative ones. Here, "positive" means that the costs of pipes were higher before optimization than after optimization.

Table 1. Wilcoxon Signed Rank Test for Cost of the Network

	Ranks	N	Mean Rank	Sum of Ranks
Pre-Cost – Post-Cost	Negative Ranks	7a	14.57	102
	Positive Ranks	23b	15.78	363
	Ties	62c		
	Total	92		

a. Precost<Postcost
b. Pre-cost>Post-cost
c. Pre-cost=Post-cost

Looking at the test statistic summary, $P=0.007 < 0.05=\alpha$, it suggest that the observed data are inconsistent with the assumption that the null hypothesis is true, that the hypothesis must be rejected. However, from 95% Monte Carlo confidence interval, the lower bound was 0 while the upper bound was 0.259. This indicates that there is no much important difference between the two costs. In summary, there is not enough evidence to reject the null hypothesis.

Pressure: A statistical comparison of the Pressures at the nodes before and after optimization indicates that the mean pressure before optimization is 12.3 while for Post optimization is 13.4. Using the null hypothesis that, “there is no difference between the two pressures before and after optimization of the water distribution network”. On this basis, the alternative hypothesis is, “there is difference between the pressure after optimization and before optimization”. From Table 3.0, it shows that the "Positive Rank" has a much greater sum than the negative ones. Here, "positive" means that the pressures at the Nodes were higher after optimization than before optimization and zero tie shows that the two pressures compare are not equivalent in any way.

Table 2. Detailed Statistical Test Results for Cost of the Network

Test Statistics[b]

	Pre-Cost – Post-Cost
Z	-2.684[a]
Asymp. Sig. (2-tailed)	0.007

a. Based on negative ranks.
b. Wilcoxon Signed Ranks Test

Table 3. Wilcoxon Signed Rank Test for pressures at the Nodes

	Ranks	N	Mean Rank	Sum of Ranks
pressure After - pressure Before	Negative Ranks	0[a]	.00	.00
	Positive Ranks	65[b]	33.00	2145.00
	Ties	0[c]		
	Total	65		

a. Pressure After<Pressure Before
b. Pressure After>Pressure Before
c. Pressure After=Pressure Before

From the test statistic summary in Table 4.0, $P=0.00 < 0.05=\alpha$, the level of significance of 0.05 (or 5%) is chosen. The p-value is less than this limit, the result is significant and it is agreed that the null hypothesis should be rejected and the alternative hypothesis-that there is a difference-is accepted.

Flow rate: The population data indicate non-parametric from the test of normality which has $P=0.001<0.05$, Using the Wilcoxon Rank test, A statistical comparison of the Flows in the pipes before and after optimization, derived from the null hypothesis, "There is no difference between the two flow rates with respect to pre and post optimization." The result of Wilcoxon Rank test (see Table 5.0) show that the sum of the "positive rank" is higher than that of the "Negative rank", it means that the flow rate before optimization is lower than the flow rate after optimization.

Table 4. Statistical Test Results for Pressures at Nodes

Test Statistics[b]

	Pressure After – Pressure Before
Z	-7.009[a]
Asymp. Sig. (2-tailed)	.000

a. Based on negative ranks.
b. Wilcoxon Signed Ranks Test

Table 5. Wilcoxon Signed Rank Test for Flow in Links

Ranks		N	Mean Rank	Sum of Ranks
After Flow – Before Flow	Negative Ranks	23[a]	36.91	849.00
	Positive Ranks	57[b]	41.95	2391.00
	Ties	12[c]		
	Total	92		

a. Flow After<Flow Before
b. Flow After>Flow Before
c. Flow After=Flow Before

Looking at the test statistic summary, $P=0.003<0.05=\alpha$, it suggest that the observed data are in consonant with the assumption that the null hypothesis is true, that the hypothesis must be rejected. This indicates that there is difference between the two flows. In summary, there is high chance to accept the alternative hypothesis, that the flow after optimization is difference with the flow before optimization.

Table 6. Statistical Test Results for Flows in Pipes

Test Statistics[b]

	AfterFlow - BeforeFlow
Z	-1.698[a]
Asymp. Sig. (2-tailed)	.003

Headloss: As regard to the null hypothesis regarding the Head loss before optimization has no difference with the Headloss after Optimization, The statistical rank test (see Table 7.0) shows that the sum of "Positive Rank" has a much greater sum than the negative ones. Its means that the Headloss in the links were higher after optimization than before optimization

Table 7. Wilcoxon Signed Rank Test for Head losses in Links

	Ranks	N	Mean Rank	Sum of Ranks
After Flow – Before Flow	Negative Ranks	26[a]	37.62	924.00
	Positive Ranks	54[b]	39.87	2316.00
	Ties	12[c]		
	Total	92		

a. Headloss After < Headloss Before
b. Headloss After > Headloss Before
c. Headloss After = Headloss Before

Observing the test statistic summary, $P=0.001<0.05=\alpha$, it suggest that the observed data are discrepant with the assumption that the null hypothesis is true, the hypothesis must be rejected. This shows that there is difference between the Headloss. In summary, the alternative hypothesis will be accepted; the Headloss after optimization is difference with the Headloss before optimization, even though the difference is insignificant.

Table 8. Statistical Test Results for Head losses in Links

Test Statistics[b]

	AfterFlow - BeforeFlow
Z	-3.518[a]
Asymp. Sig. (2-tailed)	.001

Several notes can be deduced from the analysis of result obtained after optimizing the network, this include;

1) Mathematically, the program using MATLAB, can give the corresponding diameter combinations for each link. Which means at each links in view an equivalent pipe can produce an optimum result.

2) The selected diameters and hydraulic analysis result of node and link values of the network at the end of final run of the solver. All velocity and Pressure head results are in allowable range defined by constraints

3) The Pressures (m) of the Post-Optimization hydraulic results of the Nodes are greater than that of the Pre-Optimization. This indicates that the pressure increases with increase in flow rate and decrease in pipe diameters within the nodes. The maximum pressure in the pre-optimization results is at node J2 , of value 14.79m while Node J13 have the maximum pressure of 15.71m after optimization. The minimum pressure in the Network is at J52 of value 7.59m in the former and 9.42m for the later. This shows that the minimum pressure occurs at the same node but maximum pressure shifted to another node. The statistical comparison using two related sample method of non-parametric distribution (Wilcoxon Signed Rank Test), shows a significant difference between the pressures compared. In summary the pressure increase makes the network more hydraulically efficient because for peak demand period, the pressure will be sufficient to accommodate the demand and the problem of shortage of supply is resolved. Hence, the minimum pressure of 9.42m achieved after optimization is sufficient.

4) There was a slight increase in head loss after Optimization by considering the statistical result, but from direct comparison it is

observed that the maximum head loss is at Link P95 (7.62m/km) compare to the same Link before Optimization which was 21m/km, meaning that the head loss decrease three fold at P95 and is closer to the maximum Headloss of 7.7m/km of pipes ranging from 100mm to 400mm sizes. Again the minimum head loss at link P92 was 0.05m/km after optimization while before optimization it was 0.08m/km.

5) As optimization is aimed at reducing the total cost possible, the total cost initially computed was $716501.868 for 100mm and 200mm pipe which decreased by 7.15% after optimization (as a direct comparison). Using the statistical analysis, the cost of optimized network decrease to some certain degree differs from the cost before optimization couple with improvement of the hydraulic parameters (properties) of the entire water distribution network.

4. CONCLUSIONS

Both the capital and maintenance cost of a water distribution network (not including operation cost) is enormous. As a result design Engineers are looking for new approach for the best design of water distribution networks in addition with the usual methods. In this research, a methodology is designed which uses an optimization procedure employing linear relationship of the model parameters (LOP) while the objective function ensured minimization of the capital cost of the pipes.

A designed Network of Dukku water distribution system is used as a case study. Instead of using an Extended period simulation (EPS) in running the solver, a single period simulation (SPS) is used due to the sources of the water which was mainly Bore hole.

Although the WDS is initially design with two types of pipes (200mm and 100mm pipes), and being that the minimum acceptable pipe size for any WDS is 100mm, the optimization is done using pipes of 100mm,150mm, 200mm and 250mm diameters. There is a total decreased of 7.15% (direct comparison) of the initial estimate of the pipes. But using statistical comparison it shows slight (minimal) significant differences of the cost before and after optimization. In terms of hydraulic properties the network is more efficient and pressures can accommodate peak demand at every part of the distribution network which solves the problems of water shortage in Dukku if implemented.

It is important to know that a node isolation approach can be used especially in the decision making process for improving the network if a limited amount of money is available.

ACKNOWLEDGEMENTS

This study is supported by Spicks Associates a Water Resources based Consultants. The authors wish to acknowledge the help and corporation received from the staff of the Water Resources and Environmental Engineering (ABU) Zaria in conducting this research work, and all other relevant agencies that make the research possible.

REFERENCES

1. Akdoğan, T., 2005, Design of Water distribution system by optimization using Reliability considerations (Doctoral dissertation, Middle East Technical University).

2. Burgschweiger J, Gna dig B, & Marc C., 2004, Optimization Models for Operative Planning in Drinking Water Networks. ZIB-Report, Berlin-Dahlem Germany.

3. Simon B., 2007, Closing the loop in water supply optimization. the IET Water Event, Derceto Limited, Auckland, New Zealand.

4. Mahdi M. J., 2008, Performance measurement of water distribution system (WDS).Thesis-Civil Engineering, University of Toronto, Canada.

5. Micheal H., 2012, Critical Node Analysis for Water Distribution System using flow distribution. California Polytechnic University, USA, p. 50-145.

6. Jacobs, P., & Coulter, I., 1991, Estimation of maximum cut-set size for water network failure. Journal of Water Resources Planning and Management, 117(5), 588-605.

7. Tanyimboh, T.T., Tabesh, M. & Burrows, R., 2001, Appraisal of Source head methods for calculating Reliability of Water distribution networks,

Journal of Water Resources Planning and Management, ASCE, 127(4), pp 206-213.

8. Kalungi, P., & Tanyinboh, T.T., 2003, Redundancy Model for Water distribution system, Reliability Engineering and system safety, 82(3), 275-286.

9. Klebber T. M. et al., 2003, Optimal Design of Water Distribution System by Multiobjective Evolutionary Methods. São Carlos School of Engineering, University of São Paulo, São Carlos - SP – Brazil.

10. Khalid A. M.H., 2008, Cost value function of water distribution Network, a reliability Based approached using MATLAB. An-najwah National University, Palestine.

11. Čistý M. and Bajtek Z., 2009, Optimization of the water distribution Networks with search space reduction. Slovak University of Technology, Bratislava.

12. Mays, L. W., 2000, Reliability analysis of water distribution system, ASCE, New York.

13. Zheng, Y.W., & Walski T.M., 2005. Optimizing water system improvement for a growing community. International conference of computing and control in the water industry, Exeter, UK.

14. Vasan A. & Simonovic S.P., 2010, Optimization of Water Distribution Network Design Using Differential Evolution. Journal of Water Resources Planning and Management, Vol. 136, No. 2, ASCE.

15. Deprada C (prof.), 2000, Basic concept of Optimization. Lecture note, Department of System Engineering and Automatic Control, UVA.

16. Alperovits, E., & Shamir, U., 1977, Design of optimal water distribution systems. Water Resources Research, 13(6), 885–900.

17. Quindry, G., Brill, E., & Liberman, J., 1981, Optimization of Looped water distribution systems, Journal of Environmental Engineering, ASCE, New York (NY).

18. Shamir, U., & Howard, C. D., 1985, Reliability and risk assessment for water supply systems. In Computer applications in water resources, pp. 1218-1228, ASCE.

19. Su, Y. C., Mays, L. W., Duan, N., & Lansey, K. E., 1987, Reliability-based optimization model for water distribution systems. Journal of Hydraulic Engineering, 113(12), 1539-1556.

20. Lansey, K. E., & Mays, L. W., 1989, Optimization model for water distribution system design. Journal of Hydraulic Engineering, 115(10), 1401-1418.

21. Xu, C., & Goulter, I. C., 1999, Reliability-based optimal design of water distribution networks. Journal of Water Resources Planning and Management, 125(6), 352-362.

22. Simpson, T. W., Booker, A. J., Ghosh, D., Giunta, A. A., Koch, P. N., & Yang, R. J., 2004, Approximation methods in multidisciplinary analysis and optimization: a panel discussion. Structural and multidisciplinaryoptimization, 27(5), 302-313.

23. Savic, D. A., & Walters, G. A., 1997, Genetic algorithms for least-cost design of water distribution networks. Journal of water resources planning and management, 123(2), 67-77.

24. Schaake, J. C., & Lai, D., 1969, Linear programming and dynamic programming application to water distribution network design. Hydrodynamics Laboratory, Department of Civil Engineering, Massachusetts Institute of Technology.

25. Walski, T. M., 1993, Practical aspects of providing reliability in water distribution systems. Reliability Engineering & System Safety, 42(1), 13-19.

26. Spicks Associate, 2007. Final report of Dukku water distribution system, Ahmadu Bello Way, Kaduna, Nigeria, pp.6-52.

27. Von Lücken, C., Barán, B., & Sotelo, A., 2004, Pump scheduling optimization using asynchronous parallel evolutionary algorithms. CLEI Electronic Journal,7(2).

28. Gilat, A. 2004. MATLAB: An Introduction with Applications 2nd Edition. John Wiley & Sons. ISBN 978-0-471-69420-5.

29. Xenophontos, C., 1998, A Beginners Guide to MATLAB, Technical Report, Department of Mathematics and Computer Science, Clarkson University.

30. Rossman. L., 2000, EPANET User's Manual. Cincinnati: Environmental Protection Agency.

31. Goulter, I. C., & Coals, A. V., 1986, Quantitative approaches to reliability assessment in pipe networks. Journal of Transportation Engineering.

Chapter 10

Water Demand Uncertainty: The Scaling Laws Approach

Ina Vertommen[1, 2], Roberto Magini[2], Maria da Conceição Cunha[1] and Roberto Guercio[2]

[1] Department of Civil Engineering, University of Coimbra, Coimbra, Portugal

[2] Department of Civil, Building and Environmental Engineering, La Sapienza University of Rome, Rome, Italy

1. INTRODUCTION

Water is essential to all forms of life. The development of humanity is associated to the use of water, and nowadays, the constant availability and satisfaction of water demand is a fundamental requirement in modern societies. Although water seems to be abundant on our planet, fresh water is not an inexhaustible resource and has to be managed in a rational and sustainable way. The demand for water is dynamic and influenced by various factors, from geographic, climatic and socioeconomic conditions, to cultural habits. Even within the same neighbourhood the user-specific water demand is elastic to price, condition of the water distribution system (WDS), air temperature, precipitation, and housing composition (regarding only residential demand in this case). On top of all these factors, demand varies during the day and the week.

Traditionally, for WDS modelling purposes, water demand is considered as being deterministic. This simplification worked relatively well in the past, since the major part of the studies on water demands were conducted only with the objective of quantifying global demands, both on the present and on the long-term. With the development of optimal operating schedules of supply systems, hourly water demand forecasting started to become increasingly more important. Moreover, taking in consideration all the aforementioned factors that influence water use, it is clear that demand is not deterministic, but stochastic. Thus, more recently, in order to guarantee the requested water quantities with adequate pressure and quality, the studies began to focus on instantaneous demands and their stochastic structure.

1.1. Descriptive and Predictive Models For Water Demand

The first stochastic model for (indoor) residential water demands was proposed by Buchberger and Wu [1]. According to the authors, residential water demand can be characterized by three parameters: frequency, duration and intensity, which in turn can be described by a Poisson rectangular pulse process (PRP). The adopted conceptual approach is relatively similar to basic notions of queuing theory: a busy server draws water from the system at a random, but constant, intensity and, during a random period of time. Residential demands were subdivided into deterministic and stochastic servers. Deterministic servers, including washing machines and toilets, produce pulses which are always similar. Stochastic servers, like water taps, instead produce pulses with great variability, and their duration and intensity are independent. The PRP process found to best describe water demand is non-homogeneous, i.e., when the pulse frequency is not constant in time. Different authors used real demand data to assess the adequacy of the non-homogeneous PRP model, achieving good results [2]. Moreover, the PRP model was confirmed to allow the characterization of the spatial and temporal instantaneous variability of flows in a network, unlike the traditional models that use spatial and temporal averages and neglect the instantaneous variations of demand. One drawback to the rectangular pulse based models is the fact that the total intensity is not exactly equal to the sum of the individual intensities of overlapping pulses, due to increased head loss caused by the increased flow [3]. This problem can however be solved by introducing a correction factor. The daily variability of demand represents another drawback to the PRP model, since it can invalidate the hypothesis that pulses arrive following a time dependent Poisson process [2]. One possible solution to this question is to treat the time dependent non-homogeneous process as a piecewise homogeneous process, by dividing the day into homogeneous intervals [4]. Another solution consists in using an alternative demand model: the cluster Neyman-Scott rectangular pulse model (NSRP), proposed by Alvisi [5]. The model is similar to the PRP model, but the total demand and the frequency of pulses are obtained in different ways. In the PRP model the total water demand follows a Poisson process resulting from the sum of the single-user Poisson processes, with a single arrival rate. In the NSRP model, a random number of individual demands (or elementary demands) are aggregated in demand blocks. The origin of the demand blocks is given by a Poisson process, with a certain rate between the subsequent arrivals. The temporal distance between the origins of each of the elementary demands to the origin of the demand block, follows an exponential distribution with a different rate. The variation of these parameters during the day reflects the cyclic nature of demands. A good approximation of the statistical moments for different levels of spatial and temporal aggregation was achieved; however, the variance of demand becomes underestimated for higher levels of spatial aggregation.

The aforementioned models are mainly descriptive. More recently, Blokker and Vreeburg [6] developed a predictive end-use model, based on statistical information about users and end uses, which is able to forecast water demand patterns with small temporal and spatial scales. In this model, each end-use is simulated as a rectangular pulse with specific probability distribution functions for the intensity, duration and frequency, and a given probability of use over the day. End-uses are discriminated into different types (bath, bathroom tap, dish washer, kitchen tap, shower, outside tap, washing machine, WC). The statistical distribution for the frequency of each end-use was retrieved from survey information from the Netherlands. The duration and intensity were determined, partly from the survey and partly from technical information on water-using appliances. From the retrieved information, a diurnal pattern could be built for each user. Users represent a key point in the model and are divided into groups based on household size, age, gender and occupation. Simulation results were found to be in good agreement with measured demand data. The End-Use model has also been combined with a network solver, obtaining good results for the travel times, maximum flows, velocities and pressures [7].

The PRP and the End-Use model were compared against data from Milford, Ohio. The achieved results showed that both models compare well with the measurements. The End-Use model performs better when simulating the demand patterns of a single family residence, while the PRP models describes more accurately the demand pattern of several aggregated residences [8]. The main difference between the models is the number of parameters they use: the PRP model is a relatively simple model that has only a few parameters, while the End-Use model has a large number of parameters. However, the End-Use model is very flexible towards the input parameters, which also have a clearer physical meaning and hence more intuitive to calibrate. The PRP model describes the measured flows very well. From the analytic description provided by the PRP model, a lot of mathematical deductions can be made. Thus, one can classify the PRP model as a descriptive model with a lot of potential to provide insight into some basic elements of water use, such as peak demands [9]and cross-correlations [10]. The End-Use model is a Monte Carlo type simulation that can be used as a predictive model, since it produces very realistic demand patterns. The End-Use model can be applied in scenario studies to show the result of changes in water using appliances and human behaviour. Possible improvements to the model include the incorporation of leakage, the consideration of demands as a function of the network pressure and the application of the model outside the Netherlands [11]. Li [10]studied the spatial correlation of demand series that follow PRP processes. It was verified that while time averaged demands that follow a homogeneous PRP process are uncorrelated, demands that follow a non-homogenous PRP process are correlated, and that this correlation increases with spatial and temporal aggregation. A similar conclusion about the correlation was achieved by Moughton [12]from field measurements.

1.2. Uncertainty and Reliability-Based Design Of Water Distribution Systems

The problem of WDS design consists in the definition of improvement decisions that can optimize the system given certain objectives. As aforementioned, in the earliest works regarding the optimal design of water distribution systems (WDS), input parameters, like water demand, were considered as being deterministic, often leading to under-designed networks. A robust design, allowing a system to remain feasible under a variety of values that the uncertain input parameters can assume, can only be achieved through a probabilistic approach. In a probabilistic analysis the input parameters are considered to be random variables, i.e., the single values of the parameters are replaced with statistical information that illustrates the degree of uncertainty about the true value of the parameter. The outcomes, like nodal heads, are consequently also random variables, allowing the expression of the networks' reliability.

Uncertainty in demand and pressure heads was first explicitly considered by Lansey [13]. The authors developed a single-objective chance constrained minimization problem, which was solved using the generalized reduced gradient method GRG2. The obtained results showed that higher reliability requirements were associated to higher design costs when one of the variables of the problem was uncertain.

Xu and Goulter [14] proposed an alternative method for assessing reliability in WDS. The mean values of pressure heads were obtained from the deterministic solution of the network model. The variance values were obtained using the first-order second moment method (FOSM). The probability density function (PDF) of nodal heads defined by these mean and variance values was used to estimate the reliability at each node. The approach proved to be suitable for demands with small variability. Kapelan [15] developed two new methods for the robust design of WDS: the integration method and the sampling method. The integration method consists in replacing the stochastic target robustness constraint (minimum pressure head) with a set of deterministic constraints. For that matter it is necessary to know the mean and standard deviation of the pressure heads. However, since pressure heads are dependent of the demands, it is not possible to obtain analytically the values for the standard deviations. Approximations of the values of the standard deviations are obtained by assuming the superposition principle, which makes it possible to estimate the contribution of the uncertainty in demand on the uncertainty of pressure heads. The sampling method is based on a general stochastic optimization framework, this is, a double looped process consisting on a sampling loop within an optimization loop. The optimization loop finds the optimal solution, and the sampling loop propagates the uncertainty in the input variables to the output variables, thus evaluating the potential solutions.

The aforementioned optimization problems are formulated as constrained single-objective problems, resulting in only one optimal solution (minimum cost), that provides a certain level of reliability. More recently, these optimization problems have been replaced with multi-objective problems. Babayan [16] formulated a multi-objective optimization problem considering two objectives at the same time: the minimization of the design cost and the maximization of the systems' robustness. Nodal demands and pipe roughness coefficients were assumed to be independent random variables following some PDF.

At this point, all the aforementioned models assume nodal demands as independent random variables. However, in real-life demands are most likely correlated: demands may rise and fall due to the same causes. Kapelan [17] introduced nodal demands as correlated random variables into a multi-objective optimization problem. The authors verified that the optimal design solution is more expensive when demands are correlated than the equivalent solution when demands are uncorrelated. A similar conclusion was achieved by Filion [18]. These results sustain that assuming uncorrelated demands can lead to less reliable network designs. Thus, even if increasing the complexity of optimization problems, demand correlation should always be taken into account in the design of WDS.

The robust design of WDS has gained popularity over the last years. Researchers have been focusing on methods and algorithms to solve the stochastic optimization problems, and great improvements have been made in this aspect. However, the quantification of the uncertainty itself has not been addressed. Values for the variance and correlation of nodal demands are always assumed and no attention is being paid in properly quantifying these parameters. The optimization problems could be significantly improved if more realistic values for the uncertainty would be taken into account.

This work addresses the need to understand in which measure the statistical parameters depend on the number of aggregated users and on the temporal resolution in which they are estimated. It intends to describe these dependencies through scaling laws, in order to derive the statistical properties of the total demand of a group of users from the features (mean, variance and correlation) of the demand process of a single-user. Being part of the first author's PhD research, which aims the development of descriptive and predictive models for water demand that provide insight into peak demands, extreme events and correlations at different spatial and temporal scales, these models will, in future stages, be incorporated in decision models for design purpose or scenario evaluation. Through this approach, we hope to develop more realistic and reliable WDS design and management solutions.

2. STATISTICAL CHARACTERIZATION OF WATER DEMAND

Recent studies on uncertainty in water distribution systems (WDS) refer that nodal demands are the most significant inputs in hydraulic and water quality models [19]. The variability of water demand affects the overall reliability of the model, the assessment of the spatial and temporal distributions of the pressure heads, and the evaluation of water quality along the different pipes. These uncertainties assume a different importance depending on the spatial and temporal scales that are considered when describing the network. The degree of uncertainty becomes more relevant when finer scales are reached, i.e., when small groups of users and instantaneous demands are considered. Thus, for a correct and realistic design and management, as well as simulation and performance assessment of WDS it is essential to have accurate values of water demand that take into account the variability of consumption at different scales. For that matter, the thorough description of the statistical properties of demand of the different groups of customers in the network, at specific temporal resolutions, is essential.

For a better understanding of this aspect, let us consider the distribution of the customers in a network.Figure 1 shows the network of a small town where the customers can be classified mainly as residential.

The most peripheral pipe serves the inhabitants of one single building. When moving upwards in the network, the number of customers increases reaching a maximum of 1258 customers near the tank. Obviously, as a consequence, the mean flow increases from the peripheral building to the tank. The increase of the variance of the flow is, however, less obvious. For larger networks and more densely populated towns, the difference between the number of customers that are close and far from the tank, and consequently, the variations of the mean and variance of the flow is even more pronounced.

Another important aspect when modelling a network is the choice of the adequate temporal resolution. This choice depends on the characteristics of the available measurement instruments and on the type of analysis to perform. When modelling the peripheral part of a network, characterized by a significant temporal variation of demand, it is important to adopt fine temporal resolutions, i.e., in the order of seconds. For the estimation of peak flows in design problems Tessendorff [20] suggests the use of different temporal resolutions on different sections of the network: the author suggests the use of a 15 second time interval for customer installation lines, two minutes for service lines, 15 minutes for distribution lines, and 30 minutes for mains and secondary feeders. The statistical properties of water demand are affected by the considered temporal resolution. The use of longer sampling intervals causes an inevitable loss of information about the signals, resulting in lower estimates for the variance [21, 22]. This aspect is particularly relevant at the peripheral pipes of the network that, as aforementioned, are characterized by large demand fluctuations. Therefore, understanding the spatial and temporal scaling properties of water demand is essential to build a stochastic model for water consumption.

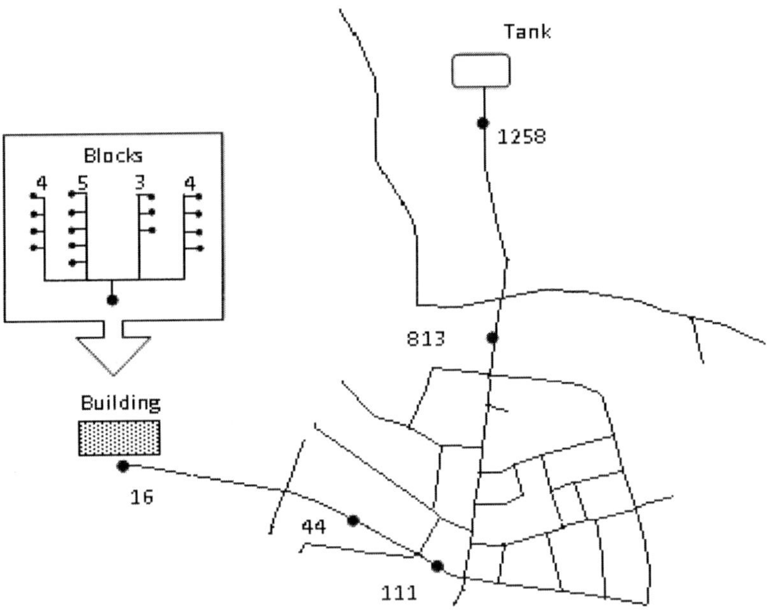

Figure 1. Spatial variability of customers in a real distribution network, from Magini *et al.* [25]. The number of customers is outlined at various locations.

Water demand can be described by a stochastic process in which q(i,t)q(i,t)represents the demand of water of the single-user iiat time instanttt. In order to estimate the statistical properties of water demand, it is necessary to have a historical series of observations, extended to sufficiently wide number of users of each type. From this data it is then possible to estimate the mean and variance of the process.

If the consumers are assumed to be of the same type, the properties of demand can be considered to be homogeneous in space, this is, they are independent of the particular consumer that is taken into consideration. Regarding the temporal variability, the stochastic process can only be assumed to be stationary in time intervals during which the mean stays constant. Once the length of this time interval, TT, is established, it is possible to determine the temporal mean, $\mu 1 \mu 1$, and variance,$\sigma 21 \sigma 12$, of the demand signal of the single-userii, as followed:

$$\mu_1 = \frac{1}{T} \int_0^T q(i,\ t)dt \tag{1}$$

$$\sigma_1^2 = \frac{1}{T} \int_0^T [q(i,\ t) - \mu_1]^2 dt \tag{2}$$

For homogeneous and stationary demands, the expected values for the mean and variance, $E[\mu 1]E\mu 1$and$E[\sigma 21]E\sigma 12$, obtained from *N* observations, provide the mean and variance of the process.

2.1. Correlation Between Consumers

The definition of the mean and variance for each type of consumer is not enough for a complete statistical characterization of demand. In order to obtain a realistic representation of the demand loads at the different nodes in a network; essential for the assessment of the network performance under conditions as close as possible to the actual working conditions, the correlation between nodal demands cannot be ignored. This correlation can be expressed through the cross-covariance and cross-correlation coefficient functions.

The cross-covariance, covAB , and cross-correlation coefficient,ρAB , between user iiof group A and user j of group B, during the observation periodTT, are expressed, respectively, as followed:

$$cov_{AB} = \frac{1}{T}\int_0^T [q_A(i_A, t) - \mu_A][q_B(j_B, t) - \mu_B]dt \tag{3}$$

$$\rho_{AB} = \frac{cov_{AB}}{\sigma_A \cdot \sigma_B} \tag{4}$$

As known, the WDS need to guarantee minimum working conditions, this is, the minimum pressure requirements have to be satisfied at each node even under maximum demand loading conditions. If all the consumers in the network are of the same type, it seems reasonable to assume a perfect correlation between demands, and to simplify the analysis of the network by assigning the same demand pattern to all the consumers. The synchronism of demands is the worst scenario that can occur on a network, causing the widest pressure fluctuations at the nodes. The assumption of a perfect correlation for design purposes results in reliable networks, but it also requires the increase of the pipe diameters, which consequently increases the networks cost. In fact, as mentioned earlier, each consumer has his own demand pattern based on specific needs and habits, without knowing what other consumers are doing at the same time. This means that demand signals in real networks are correlated, but are not synchronous. Thus, in order to obtain the optimum design of a network, it is essential to estimate the accurate level of correlation between the consumers. On the other hand, to estimate accurately the spatial correlation between demands, it is necessary to collect and analyse historical series, resulting in additional costs in the design phase. However, these additional costs will most certainly be compensated by the achieved reduction of the following construction costs.

2.2. The Scaling Laws Approach In Modelling Water Demand Uncertainty

Water demand uncertainty is made of both aleatory or inherent uncertainty, due to the natural and unpredictable variability of demand in space and time, and epistemic or internal uncertainty, due to a lack of knowledge about it. Hutton [23] distinguishes epistemic uncertainty in two types. The first type concerns the nature of the demand patterns, and the lack of knowledge about this variability when modelling WDS both in time and space. This uncertainty is defined as 'two-dimensional' uncertainty since it is composed by both aleatory and epistemic uncertainty. It can be reduced with extended and expensive spatial and temporal data collection or through the employment of descriptive and predictive water demand models. The second type of epistemic uncertainty takes the spatial allocation of water demand into account when modelling WDS [24].

Dealing with the 'two-dimensional' uncertainty when modelling WDS, requires not only a complete statistical characterization of demand variability, but also the determination of the correlation among the different users and groups of users. The natural variability of demand can be expressed using probability density functions (PDF). A PDF is characterized by its shape (e.g. normal, exponential, gamma, among others) and by specific parameters like the population mean and variance. Thus, in order to represent uncertain water demand using a PDF, it is necessary to identify and estimate the values of these parameters. The consideration of different spatial and temporal aggregation levels induces changes in the PDF parameters, often leading to a reduction of the uncertainty. The auto-correlation and cross-correlation that characterize the water demand signals affect the extent to which the PDF parameters vary, and can introduce an additional sensitivity to the specific period of observation in question.

In order to understand the effects of spatial aggregation and sampling intervals on the statistical properties of demand, it is possible to develop analytical expressions for the moments (mean, variance, cross-covariance and cross-correlation coefficient) of demand time series, at a fixed time sampling frequency Δt, of nnaggregated users as a function of the moments of the single-user series sampled in the observation period *T*. These expressions are referred to as *"Scaling Laws"*, and can be expressed as:

$$E[m_{\Delta t,T}(n)] = E[m_{\Delta t,T}] \cdot n^{\alpha} \cdot f(\Delta t,\, T) \tag{5}$$

WhereE[mΔt,T(n)]EmΔt,T(n) is the expected value of the moment mmfor nnusers for the time intervalTT;E[mΔt,T]EmΔt,T is the expected value of the moment mmfor the single-user for the same time interval; ααis the exponent of the scaling law; and f(Δt,T)f(Δt,T)is a function that expresses the influence of both sampling rate and observation period.

The development of the scaling laws is based on the assumption that the demand can be described by a homogeneous and stationary process, which implies that the nnaggregated users are of the same type (residential, commercial, industrial, etc.), and that the statistical properties of demand, mean and variance, can be assumed constant in time. The scaling laws for the mean, variance, and lag1 covariance were derived by Magini [25]. The expected value of the total mean demand q(n,t)q(n,t)can be expressed as followed:

$$E[\mu_{\Delta t,T}(n)]=E\left[\frac{1}{T}\int_0^T\sum_{j=1}^{n} q(j,\tau)d\tau\right]=\frac{1}{T}\int_0^T\sum_{j=1}^{n} E[q(j,\tau)]d\tau=n\cdot E[\mu_1] \tag{6}$$

Where E[μ1] Eμ1 is the expected demand value for the single user or 'unit mean'. This expression shows that the mean demand increases linearly with the number of users according to a factor of proportionality equal to the expected value of the single user and is independent of the sampling rate and observation period.

In order to estimate the expected value of the demand variance it is necessary to consider the covariance function cov(s,τ)cov(s,τ)of the single-user demand at the spatial and temporal lags, s=j1−j2s=j1-j2andτ=τ1−τ2τ=τ1-τ2, respectively. The following expression is obtained (see [26] for the mathematical passages):

$$\begin{aligned}E[\sigma^2_{\Delta t,T}(n)]&=\frac{1}{T^2}\int_0^T\int_0^T\sum_{i_1=1}^{n}\sum_{i_2=1}^{n}[\mathrm{cov}_{\Delta t}(s,0)-\mathrm{cov}_{\Delta t}(s,\tau)]d\tau_1 d\tau_2\\&=\sum_{i_1=1}^{n}\sum_{i_2=1}^{n}\left[\mathrm{cov}_{\Delta t}(s,0)-\frac{1}{T^2}\int_0^T\int_0^T\mathrm{cov}_{\Delta t}(s,\tau)d\tau_1 d\tau_2\right]\end{aligned} \tag{7}$$

Where covΔt(s,0) is the covariance function at lags=0s=0, and covΔt(s,τ) is the space-time covariance function. This expression shows that the expected value for the sample variance of the nn-users process depends on the correlation structure of the single-user demands. The term 1T2∫T0∫T0covΔt(s,τ)dτ1dτ2 decreases as the period of observation TTincreases, becoming negligible whenT>>θT>>θ, being θ θ a parameter, connected to the cross-correlation of the demands and similar to the scale of fluctuation for the auto-correlation of a single signal.

The term covΔt(s,0)covΔts,0 is independent from τ1τ1 andτ2τ2, and assumes the following values:

$$cov_{\Delta t}(s,0)=\begin{cases}cov_{\Delta t}(s) & j_1\neq j_2\\ \sigma^2_{1,\Delta t} & j_1=j_2\end{cases} \tag{8}$$

Where covΔt(s)covΔts is the spatial cross-covariance between different single-user demands, and σ21,Δtσ1,Δt2 the variance of the single user. For large values ofTT, equation (7) can be simplified into:

$$E[\sigma^2_{\Delta t,T}(n)] = \sum_{j_1=1}^{n}\sum_{j_2=1}^{n} cov_{\Delta t}(s,0) = \sum_{j_1=1}^{n}\sum_{j_2=j_1} cov_{\Delta t}(s,0) + \sum_{j_1=1}^{n}\sum_{j_2\neq j_1} cov_{\Delta t}(s,0)$$
$$= n \cdot \sigma^2_{1,\Delta t} + n \cdot (n-1) \cdot cov_{\Delta t}(s) \tag{9}$$

This equation represents the scaling law for the variance, neglecting the bias that can be caused when using small the demand series (short observation periods).

Introducing the Pearson cross-correlation coefficient given byρ=covxyσxσyρ=covxyσxσy, and considering ρΔtρ as the cross-correlation coefficient between each couple of single-user demands, the spatial covariance can be expressed ascovΔt=ρΔTσ21,ΔtcovΔt=ρΔTσ1,Δt2, and Equation (9) becomes:

$$E[\sigma^2_{\Delta t}(n)] = n^2 \cdot \rho_{\Delta t} \cdot \sigma^2_{1,\Delta t} + n \cdot [1 - \rho_{\Delta t}] \cdot \sigma^2_{1,\Delta t} \tag{10}$$

If demands are perfectly correlated in space then ρΔtρΔt is equal to one, and equation (10) is simplified into:

$$E[\sigma^2_{\Delta t}(n)] = n^2 \cdot \sigma^2_{1,\Delta t} \tag{11}$$

If demands are uncorrelated in space then ρΔtρΔt is equal to zero, and equation (9) is simplified into:

$$E[\sigma^2_{\Delta t}(n)] = n \cdot \sigma^2_{1,\Delta t} \tag{12}$$

Since the cross-correlation coefficient can assume values between 00 and11, equations (10) and (11) represent the maximum and minimum expected values for the variance. Equation (9) can be simplified into a more generic form given by:

$$E[\sigma^2_{\Delta t}(n)] = n^{\alpha} \cdot \sigma^2_{1,\Delta t} \tag{13}$$

Where1≤α≤2.

In conclusion, it can be stated that the variance in the consumption signal of a group of usersnn, homogeneous in type, is proportional to the mean variance of the single-user according to an exponent, which varies between 1 and 2. The value of the scaling exponent depends on the structure of the spatial correlation, i.e., the correlation that exists between the different consumptions during the observation period: if demands are uncorrelated in space, the scaling law is linear, if demands are perfectly correlated in space, the scaling law is quadratic.

The variance function, γ(Δt), measures the reduction of the variance of the instantaneous signal when the sampling interval ΔtΔtincreases [27], as followed:

$$\sigma_{1,\Delta t}^2 = \sigma_1^2 \cdot \gamma(\Delta t) \tag{14}$$

Where σ21σ12 is the variance of the instantaneous signal for the single user. Introducing the variance function in equation (13), the following is obtained:

$$E[\sigma_{\Delta t}^2(n)] = n^{\alpha} \cdot \sigma_1^2 \cdot \varphi(\Delta t) \tag{15}$$

Similarly, the expected value of the cross-covariance is given by:

$$E[cov_{AB,\Delta t}(n_a, n_b)] = \frac{1}{T^2}\int_0^T\int_0^T \sum_{i=1}^{n_a}\sum_{j=1}^{n_b} [cov_{AB,\Delta t}(s,0) - \mathrm{cov}_{AB,\Delta t}(s,\tau)]d\tau_1 d\tau_2$$
$$= \sum_{i=1}^{n}\sum_{j=1}^{n}\left[\mathrm{cov}_{AB,\Delta t}(s,0) - \frac{1}{T^2}\int_0^T\int_0^T \mathrm{cov}_{AB,\Delta t}(s,\tau)d\tau_1 d\tau_2\right] \tag{16}$$

Neglecting the term 1T2∫T0∫T0covAB,Δt(s,τ)dτ1dτ2,t ,the expected value of the cross-covariance between the demands of nana aggregated users of group A and nbnb aggregated users of group B is given by:

$$E[cov_{AB,\Delta t}(n_a, n_b)] = n_a \cdot n_b \cdot E[\rho_{ab,\Delta t}] \cdot E[\sigma_{a,\Delta t}] \cdot E[\sigma_{b,\Delta t}] \tag{17}$$

Where, E[ρab,Δt] Eρab,Δt is the expected Pearson cross-correlation coefficient between the single-user demands of the two groups; and σa,Δtσa,Δt and σb,Δtσb,Δtare the standard deviations of the single-user demands of groups AAandBB, respectively, at the sampling rateΔt. The expected value of the cross-covariance increases according to the product between the number of users of each group. In the particular case in which both groups have the same statistical properties, i.e., they belong to the same process, and assuming thatna=nbna=nb, the scaling law of the cross-covariance becomes quadratic.

As a consequence, the expected value of the Pearson cross-correlation coefficient between the demands of nana aggregated users of group A and nbnb aggregated users of group B, is given by:

$$E[\rho_{AB,\Delta t}(n_a, n_b)] = \frac{E\left[Cov_{AB,\Delta t}(n_a, n_b)\right]}{E\left[\sigma_{A,\Delta t}(n_a)\right] \cdot E\left[\sigma_{B,\Delta t}(n_b)\right]} = \frac{n_a \cdot n_b \cdot E\left[\rho_{ab,\Delta t}\right]}{\sqrt{n_a\left(1 + E\left[\rho_{a,\Delta t}\right] \cdot \left[n_a - 1\right]\right)} \cdot \sqrt{n_b\left(1 + E\left[\rho_{b,\Delta t}\right] \cdot \left[n_b - 1\right]\right)}}$$

(18)

This equation shows that this coefficient depends separately on the spatial aggregation levels of each group, n_a and n_b, and not only on their product as happens for the cross-covariance. If $n_a = n_b = n$equation [18] becomes:

$$E[\rho_{AB,\Delta t}(n)] = \frac{n \cdot E\left[\rho_{ab,\Delta t}\right]}{\sqrt{\left[1 + (n - 1) \cdot E\left[\rho_{a,\Delta t}\right]\right]\left[1 + (n - 1) \cdot E\left[\rho_{b,\Delta t}\right]\right]}}$$

(19)

From equation (19) it is possible to observe that the expected value E[ρAB,Δt(na,nb)]EρAB,Δtna,nbincreases with the number of users, nanaandnbnb, reaching the following limit value:

$$E[\rho_{AB,\Delta t}(n_a, n_b)] = \lim_{\substack{n_a \to \infty \\ n_b \to \infty}} E[\rho_{AB,\Delta t}(n_a, n_b)] = \frac{E\left[\rho_{ab,\Delta t}\right]}{\sqrt{\left(E\left[\rho_{a,\Delta t}\right] \cdot E\left[\rho_{b,\Delta t}\right]\right)}}$$

(20)

Since by definitionE[ρab,Δt]≤1E[ρab,Δt]≤1, the maximum value that the expected value of the cross-correlation coefficient between the single-user demands of group AAand BBcan assume is:

$$E[\rho_{AB,\Delta t}]_{max} = \sqrt{\left(E[\rho_{a,\Delta t}] \cdot E[\rho_{b,\Delta t}]\right)}$$

(21)

From equation (21) it is also possible to observe that the Pearson cross-correlation coefficient between the nana aggregated users of group AAand the nbnb aggregated users of group BBdepends on both the cross-correlations inside each group and the cross-correlation between the groups. Therefore, it seems interesting to investigate the way in which these two aspects, one at a time, affect the expected value of the cross-correlation when the number of aggregated users increases and for a fixed sampling rateΔtΔt. In order to do so let us first consider a fixed value of ρabρab and varying values of ρaρa andρbρb. Figure 2shows the graphical results for ρab=0.1ρab=0.1 and different pairs of ρaρa andρbρb.

As expected, all the curves have a common starting point, since ρabρab is fixed. According to equation (19) a gradual flattening of the curves and a reduction of the shape ratio ρAB,lim/ρabρAB,lim/ρabcan be noticed when the product ρa·ρbρa·ρb increases. Let us now consider a different case in which ρaρa and ρbρb are fixed and ρabρabvaries. The results are shown graphically in figure 3. The curves have now different starting points and equal shape ratiosρAB,lim/ρabρAB,lim/ρab. Increasing ρabρab produces only an upward shift of the curves, extending their transient.

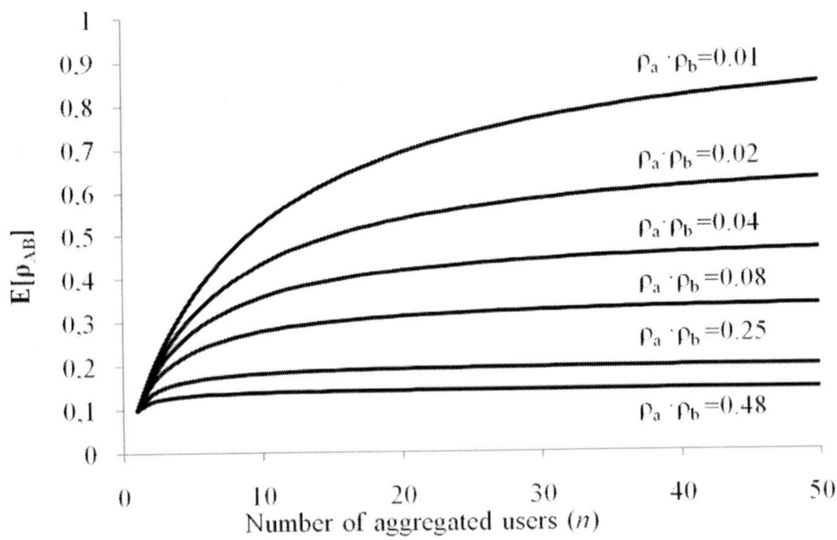

Figure 2. Scaling laws ofE[ρAB(n)]EρAB(n), for different values ofρa·ρb.

In the particular case in which both groups of users have the same statistical properties, i.e., they belong to the same process, and assumingna=nb=nna=nb=n, the scaling law for the cross-correlation coefficient, considering no differences in the sampling time intervals, is:

$$E[\rho_{AB,\Delta t}(n)] = \frac{n \cdot E[\rho_{\Delta t}]}{1 + (n - 1) \cdot E[\rho_{\Delta t}]} \tag{22}$$

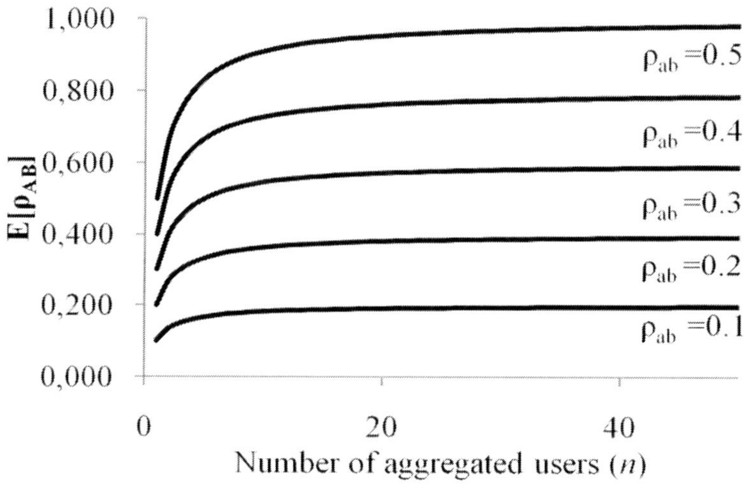

Figure 3. Scaling laws ofE[ρAB(n)]EρAB(n), for different values ofρabρab.

From equation (22) it is clear that the cross-correlation coefficientincreases with the number of aggregated users, tending to one. This limit value is reached as sooner as the cross-correlation coefficient, E[ρΔt]EρΔt, between the single-user demands is higher.

3. VALIDATION OF THE ANALYTICAL EXPRESSIONS

3.1. Synthetically Generated Signals: Scaling Laws For The Mean And The Variance

In order to confirm the analytical development reported in the previous paragraph, the scaling laws were derived for groups of synthetically and simultaneously generated consumption signals. At this aim the Multivariate Streamflow model [28], with a normal probability distribution, was used. Each group was assumed to contain 300 consumption signals with 3600 realizations each, distinguished by different values of the cross-correlation coefficient between them. The correctness of the procedure used to generate each demand series was tested by checking that the mean, the variance and the cross-correlation coefficient of the generated signals were equal to the input parameters of the model. Only little differences were observed (Table 1), which are explained due to the fact that the generated demand series are realizations of a stochastic process and, consequently, their moments necessarily differ from the theoretical ones.

Once the single consumption signals of each group were generated, they were aggregated randomly selecting one at a time, until a maximum of 100 aggregated consumption signals was reached. The first and second order moments, mean and variance, were calculated for each aggregation level. In order to obtain a result as general as possible, the same procedure has been repeated 50 times, aggregating each time different users [25]. The obtained results are summarized in Table 1 and 2, with reference to equation 5.

Results confirm the linear scaling for the first order moment and show that the variance increases with the spatial aggregation level according to an exponent that varies between 1 and 2. In theory, for spatially uncorrelated demands the scaling laws is linear and for perfectly correlated demands the scaling law is quadratic.

3.2. Synthetically Generated Signals: Scaling Laws For The Cross-Covariance

In this case pairs of aggregated consumption series, A and B, were obtained by randomly selecting among pairs of the previously generated groups of signals. Different values of the productna·nbna·nb, wherenana is the number of signals in group A and nbnb the same number in group B, were considered, up to the maximum valuena·nb=500na·nb=500. Each aggregation process was characterized by the cross-correlation value between the single signals in the same group and the cross-correlation value between the single signals of the two native groups. The cross-covariance was computed for the different aggregation levels and the scaling law were derived for each process. The results are summarized in Table 3 with reference to equation 17, considering Coeff=E[ρab,ΔT]·E[σa,ΔT]·E[σb,ΔT] and α as the exponent of the product na·nb.

Table 1. Theoretical and experimental values of the scaling law for the first order moment for different values of the cross-correlation.

Cross-correlation	$E[m_T]$		α	
coefficient	Theoretical	Experimental	Theoretical	Experimental
0	0.70	0.7003	1.00	0.9996
0.001	0.70	0.7017	1.00	0.9993
0.010	0.70	0.6971	1.00	1.0001
0.025	0.70	0.7020	1.00	0.9989
0.050	0.70	0.7096	1.00	1.0004
0.10	0.70	0.7063	1.00	1.0009
0.20	0.70	0.7086	1.00	0.9994
0.30	0.70	0.7032	1.00	1.0008
0.40	0.70	0.6923	1.00	1.0003
0.50	0.70	0.6942	1.00	0.9985
0.60	0.70	0.6857	1.00	1.0002
0.70	0.70	0.6874	1.00	1.0011
0.80	0.70	0.6852	1.00	0.9998
0.90	0.70	0.6789	1.00	1.0009
0.99	0.70	0.7050	1.00	0.9997

Table 2. Experimental values of the scaling law for the second order moment for different values of the cross-correlation.

Cross-correlation coefficient	$E[m_T]$	α
0	3.9808	1.0004
0.001	3.5205	1.0541
0.010	1.9864	1.3079
0.025	1.4498	1.4984
0.050	1.2702	1.6403
0.10	1.2940	1.7686
0.20	1.5621	1.8675
0.30	1.8879	1.9139
0.40	2.1713	1.9379
0.50	2.4485	1.9570
0.60	2.7685	1.9692
0.70	3.0803	1.9804
0.80	3.4051	1.9888
0.90	3.6567	1.9945
0.99	3.9960	1.9985

Results confirm that α α is always equal to one. However, in this case the scaling does not consider the number of aggregated users, but their product, and thus the law is not linear but quadratic. A similar approach was also applied in the particular case in whichρa=ρb=ρab, andσa=σb, that is, when all the consumptions are homogeneous, and withna=nb.

Table 3. Theoretical and experimental values of the scaling law for the cross-covariance for different values of the cross-correlation coefficients in, *ρa* and *ρb*, and between A,B, *ρab*.

Cross-correlation coefficient			Coeff		α	
ρ_a	ρ_b	ρ_{ab}	theoretical	experimental	theoretical	experimental
0.10	0.10	0.10	0.3754	0.3747	1.00	0.9998
0.20	0.10	0.10	0.3880	0.3907	1.00	0.9982
0.40	0.10	0.10	0.3676	0.3624	1.00	1.0020
0.40	0.20	0.10	0.3463	0.3440	1.00	1.0003
0.80	0.60	0.10	0.3939	0.3908	1.00	1.0009
0.50	0.50	0.10	0.3545	0.3541	1.00	0.9993
0.50	0.50	0.20	0.7449	0.7401	1.00	1.0007
0.50	0.50	0.30	1.0999	1.0923	1.00	1.0004
0.50	0.50	0.40	1.4643	1.4605	1.00	0.9999
0.50	0.50	0.50	1.8408	1.8504	1.00	0.9986

Table 4. Theoretical and experimental values of the scaling law for the cross-covariance between homogeneous groups of consumptions and different values of the cross-correlation coefficient.

Cross-correlation	Coeff		α	
coefficient	theoretical	experimental	theoretical	experimental
0.10	0.40	0.4000	2.00	1.9998
0.20	0.80	0.7914	2.00	1.9997
0.30	1.20	1.2160	2.00	2.0008
0.40	1.60	1.6009	2.00	1.9985
0.50	2.00	2.0059	2.00	1.9992
0.60	2.40	2.3955	2.00	2.0003
0.70	2.80	2.7934	2.00	2.0000
0.80	3.20	3.2043	2.00	1.9999
0.90	3.60	3.6057	2.00	1.9999
0.99	3.96	3.9408	2.00	2.0003

Equation 17 then becomes $E[cov_{AB,T}(n_a, n_b)] = n^2 \cdot E[\rho_{ab,\Delta t}] \cdot E[\sigma^2_{\Delta t}]$. The results for different values of the cross-correlation coefficient are described in Table 4. They confirm the theoretical quadratic scaling for cross-covariance.

3.3. Real Consumption Data: Scaling Laws For The Mean And The Variance

The parameters of the scaling laws were also derived for a set of real demand data. The indoor water uses demand series of 82 single-family homes, with a total of 177 inhabitants, in a building belonging to the IIACP (Italian Association of Council Houses) in the town of Latina were considered [29, 30]. The apartments are inhabited by single-income families, belonging to the same low socioeconomic class.The daily demand series of four different days (4 consecutive Mondays) of the 82 users were considered [25]. For each user the different days of consumptions can be considered different realizations of the same stochastic process. In this way the number of customers was artificially extended to about 300, preserving at the same time the homogeneity of the sample. The temporal resolution of each time series is 1 second.

Table 5. Estimated parameters of the scaling laws for the experimental data set of Latina (see [25]).

Time	$E[\mu_1]$	$E[\sigma^2{}_1]$	a_{var}
	(L/min)	*(L/min)²*	-
6-7	0.468	1.994	1.2288
7-8	1.066	6.678	1.114
8-9	0.988	7.401	1.0435
9-10	0.891	6.205	1.0756
10-11	0.735	4.336	1.113
11-12	0.791	4.782	1.089
12-13	0.68	4.452	1.092
13-14	0.807	5.322	1.065
14-15	0.827	5.338	1.0688
15-16	0.704	3.857	1.1311
16-17	0.512	2.266	1.1739
17-18	0.634	3.112	1.1666
18-19	0.667	3.594	1.1412
19-20	0.707	5.445	1.0384
20-21	0.68	3.702	1.1253
21-22	0.635	3.412	1.099
22-23	0.397	1.958	1.0771
mean	0.717	4.344	1.1084
confidence limits 95%	0.082	0.759	0.024

Time	$E[\mu_1]$ (L/min)	$E[\sigma^2_1]$ $(L/min)^2$	α_{var} -
6-7	0.468	1.994	1.2288
7-8	1.066	6.678	1.114
8-9	0.988	7.401	1.0435
9-10	0.891	6.205	1.0756
10-11	0.735	4.336	1.113
11-12	0.791	4.782	1.089
12-13	0.68	4.452	1.092
13-14	0.807	5.322	1.065
14-15	0.827	5.338	1.0688
15-16	0.704	3.857	1.1311
16-17	0.512	2.266	1.1739
17-18	0.634	3.112	1.1666
18-19	0.667	3.594	1.1412
19-20	0.707	5.445	1.0384
20-21	0.68	3.702	1.1253
21-22	0.635	3.412	1.099
22-23	0.397	1.958	1.0771
mean	0.717	4.344	1.1084
confidence limits 95%	0.082	0.759	0.024

The series were divided into time periods of 1 hour to guarantee the stationarity of the process. InTable 5 the estimated values of the expected values of the mean and the variance of the unit user and the exponent α for the scaling law of the variance are reported. The same exponents for the mean were always trivially equal to 1. In these results the first six hours of the day and the last one were excluded because, during the night hours consumptions are very small and therefore their statistics have a poor significance. It was observed that the mean scales linearly with the number of customers. Differently, the variance shows a slight non-linearity with the number of users. It must be underlined that the average daily value of the exponent α is 1.1, showing that there is a very weak correlation between the considered users.

3.4. Real Consumption Data: Scaling Laws For The Cross-Covariance Andcross-Correlation Coefficient

Considering the consumption signals belonging to homogeneous users, equation 23 is valid and a quadratic scaling law for the cross-covariance should be expected. This behaviour was confirmed by the measured data for all the time intervals considered. In Figure 4 the scaling law of the consumption signals between 11:00 and 12:00 am is graphically reported.

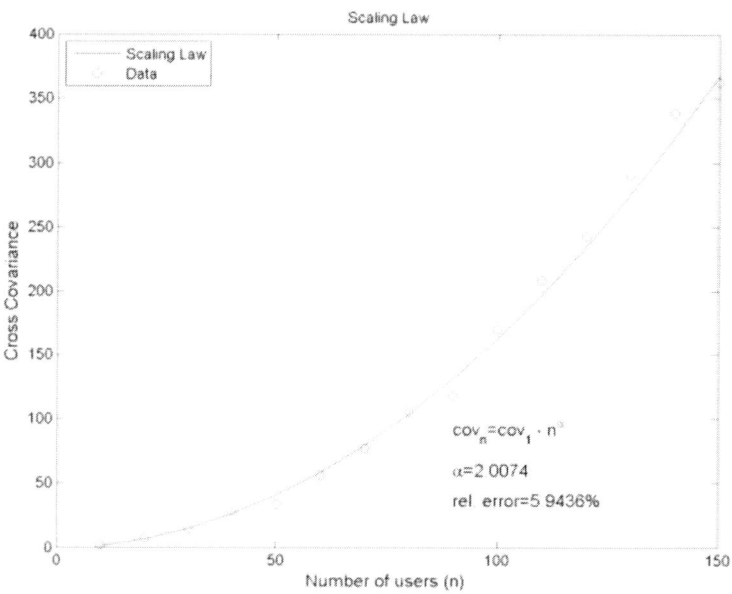

Figure 4. Scaling law for the cross-covariance between 11:00 and 12:00am.

The obtained cross-correlation coefficient between the single user signals was low, being always less than 0.05, but increased noticeably when the number of aggregated users increased, as expected according to equation 22. For groups of 150 aggregated users the cross-correlation coefficient reached the values shown in Table 6. These results enhance the importance of evaluating the cross-correlation degree at different levels of spatial aggregation. Even if the cross-correlation between single-user demand signals is relatively low and less likely to significantly affect the performance of a network, it can largely increase with the spatial aggregation of users, becoming not negligible at those larger scales.

Table 6. Estimated values of the mean cross-correlation coefficients between groups of 150 aggregated user from the experimental data set of Latina.

Time	$\rho_{150users}$	Time	$\rho_{150users}$
0-1	0.56	13-14	0.49
1-2	0.6	14-15	0.5
3-4	0.5	15-16	0.42
4-5	0.58	16-17	0.49
5-6	0.36	17-18	0.5
6-7	0.48	18-19	0.39
7-8	0.71	19-20	0.51
8-9	0.61	20-21	0.52
9-10	0.39	21-22	0.39
10-11	0.46	22-23	0.47
11-12	0.57	23-24	0.61
12-13	0.48	-	-

4. STOCHASTIC SIMULATION OF A NETWORK

To illustrate the effect of the uncertainty of water demands on the performance of a network, particularly, the effect of the level of correlation between consumptions on the outcome pressure heads, a simple network simulation was performed. The water distribution network of Hanoi (Fujiwara and Khang, 1990) was considered for this matter (Figure 5).

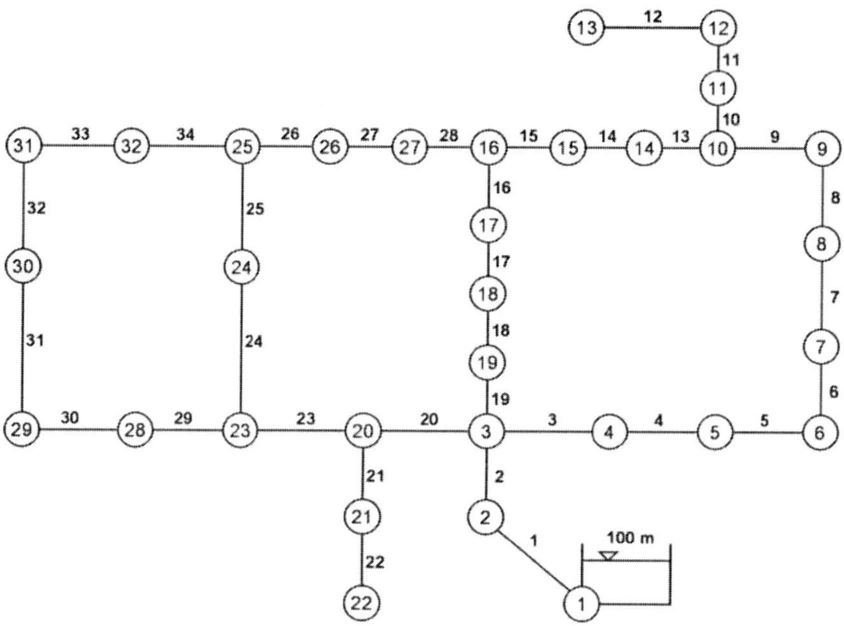

Figure 5. Water distribution network of Hanoi [31].

The data for the Hanoi network were taken from the literature (Fujiwara and Khang, 1990), and the pipe diameters were assumed to be the ones obtained by Cunha and Sousa (2001). The demand data from the literature was used to estimate the number of users at each node, assuming a single-user mean demand of 0.002 l/s. All the users in the network were assumed to be residential and having the same characteristics. The standard deviation of demand was assumed to be 0.06 l/s. The Multivariate Streamflow model [28] was used to generate synthetic stochastic demands with different levels of cross-correlation between the single-users. The nodal demands were then introduced in the network and the performance of the network was simulated using EPANET [32]. For each considered degree of cross-correlation between demands, 100 simulations were performed, resulting in series of pressure heads for each node and for each correlation level.

The first aspect that emerges from the simulations, is that the number of nodes that fail, i.e., which do not satisfy the minimum pressure requirements, increase when the cross-correlation degree increases. Higher correlations imply more synchronous consumptions, leading to pressure failures. Figure 6illustrates this result.

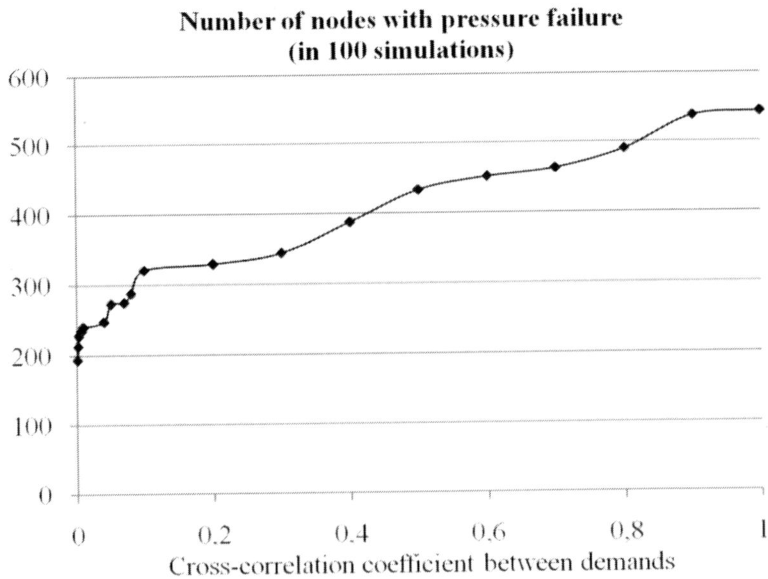

Figure 6. Total number of nodes that do not satisfy minimum pressure requirements in 100 simulations.

Observing Figure 6 it is clear that the cross-correlation between demands significantly affects the outcome pressure heads. The number of nodes that do not satisfy the minimum pressure requirements in the network increase from 194 nodes (total nodes in the network that fail in 100 simulations) when the cross-correlation between demands is equal to 0.001, to a total of 543 nodes when the cross-correlation between demands is 0.999. In other words, the probability of failure increases from 6.3% to 17.5% between the minimum and maximum levels of cross-correlation that were considered.

Another aspect that emerges from the simulations is the increase of standard deviation of the pressure heads at each node of the network, which is illustrated in Figure 7.

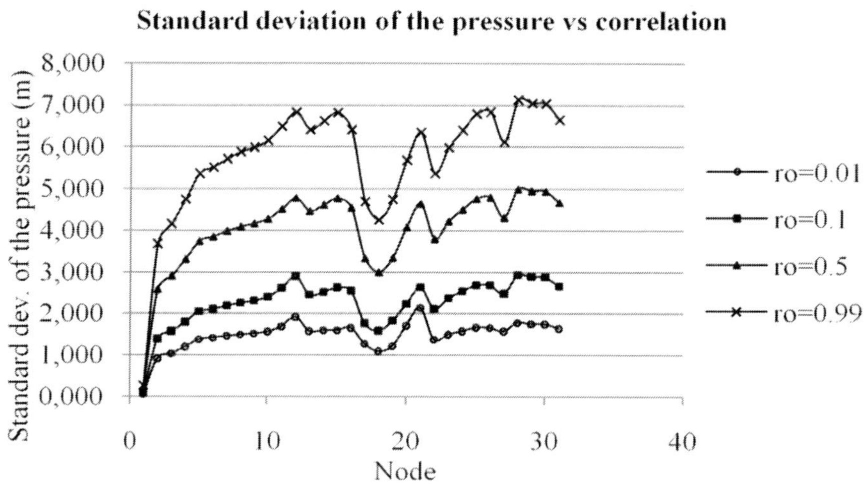

Figure 7. Standard deviation of the pressure heads *vs* the cross-correlation coefficient between the demands.

The standard deviation of the pressure head verified at each node increases when the cross-correlation between demands increases. The average standard deviation of the pressure heads along the network when the cross-correlation between demands is equal to 0.01 is 1.35m, while the average standard deviation of the pressure heads when the cross-correlation is 0.99, is 5.75m. This means that the cross-correlation increases from 0.01 to 0.99 the standard deviation of the pressure heads increases more than 4 times.

The obtained results clearly show that the level of cross-correlation between demands significantly affects the performance of a network and should, therefore, not be ignored when designing and managing WDS.

5. CONCLUSIONS

Understanding and modelling the stochastic nature of water demand represents a challenging field for researchers. Stochastic modelling faces difficulties like scarce availability of data for calibration purposes, high computational efforts associated to simulations, and the complexity of the problem itself. Moreover, the statistical properties of water demand change with the spatial and temporal scales that are used, which makes it even more difficult to accurately model the stochastic structure of demand. The proposed scaling laws represent a step forward in understanding the relation between the parameters that describe probabilistic demands and the spatial and temporal scales in which demands are measured and in which they should be modelled for WDS design or management purposes. The use of scaling laws allow a more accurate quantification of the statistical parameters, like variance and correlation, based on the real demand patterns, number of users at each node and the sampling time that is used. The scaling laws also allow to easily change the scale of the problem, since the statistical parameters and levels of uncertainty can be derived for any desired time or spatial scale.

The scaling laws were derived analytically and validated using synthetically generated stochastic demands and real demand data from Latina, Italy. A good agreement was found between the theoretical expressions, the synthetic demand data and the real demand data. Results show that the mean increases linearly with the number of aggregated users. The variance increases with spatial aggregation according to an exponent that varies between 1 and 2. In theory, for spatially uncorrelated demands the scaling laws is linear and for perfectly correlated demands the scaling law is quadratic. This aspect is clearly verified by the synthetic data. The scaling law for the covariance between 2 groups of users increases according to the product between the numbers of users in each group. The cross-correlation coefficient depends separately on the number of users in each group, and increases towards a limit value. Even for small values of cross-correlation between single-user demands, this parameter cannot be ignored since it significantly increases with the aggregation of consumers.

The performed network simulation considering stochastic demands with different pre-defined levels of correlation show a clear influence of the degree of correlation on the outcome pressure heads: higher levels of correlation lead to larger fluctuations of the pressure heads and to more frequent pressure failures. At this point, the stochastic correlated demands were only used for simulation purposes. However, in future work a similar approach, can be used for design and management purposes. The consideration of correlated stochastic demands will result in more realistic and reliable water distribution networks.

ACKNOWLEDGEMENTS

The participation of the first author in the study has been supported by Fundação para a Ciência e Tecnologia through Grant SFRH/BD/65842/2009.

REFERENCES

1. S. G. Buchberger, L. Wu, 1995 Model for instantaneous residential water demands Journal of Hydraulic Engineering, ASCE 121 3 232 246

2. S. G. Buchberger, G. J. Wells, 1996 Intensity, Duration, and Frequency of Residential Water Demands Journal of Water Resources Planning and Management, ASCE 122 1 11 19

3. V. J. Garcia, R. Garcia-Bartual, E. Cabrera, F. Arregui, J. Garcia-Serra, 2004 Stochastic Model to Evaluate Residential Water Demands Water Resources Planning and Management, ASCE 130 5 386 394

4. S. G. Buchberger, Y.. Lee, 1999 Evidence supporting the Poisson pulse hypothesis for residential water demands Water Industry systems: Modelling and optimization applications 215 227

5. S. Alvisi, M. Franchini, A. Marinelli, 2003 A stochastic model for representing drinking water demand at resindential level Water Resources Management 17 197 222

6. E. J. M. Blokker, J. H. G. Vreeburg, 2005 Monte Carlo Simulation of Residential Water Demand: A Stochastic End-Use Model Impacts of

Global Climate Change, 2005 World Water and Environmental Resources Congress Anchorage, Alaska: American Society of Civil Engineers

7. E. J. M. Blokker, J. H. G. Vreeburg, A. J. Vogelaar, 2006 Combining the probabilistic demand model SIMDEUM with a network model Water Distribution System Analysis 8th Annual Water Distribution Systems Analysis Symposium Cincinnati, Ohio, USA

8. E. J. M. Blokker, S. G. Buchberger, J. H. G. Vreeburg, J. C. van Dijk, 2008 Comparison of water demand models: PRP and SIMDEUM applied to Milford, Ohio, data WDSA Kruger Park, South Africa

9. S. G. Buchberger, E. J. M. Blokker, J. H. G. Vreeburg, 2008 Sizes for Self-Cleaning Pipes in Municipal Water Supply Systems WDSA Kruger Park, South Africa

10. Z. Li, S. G. Buchberger, D. Boccelli, Y. Filion, 2007 Spatial correlation analysis of stochastic residential water demands Water Management Chalenges in Global Change London: Taylor & Francis Group

11. E. J. M. Blokker, J. H. G. Vreeburg, J. C. van Dijk, 2010 Simulating Residential Water Demand with a Stochastic End-Use Model Journal of Water Resources Planning and Management, ASCE 136 1 19 26

12. L. J. Moughton, S. G. Buchberger, D. L. Boccelli, Y. Filion, B. W. Karney, 2006 Effect of time step and data aggregation on cross correlation of residential demands Cincinnati, Ohio, USA

13. K. E. Lansey, L. W. Mays, 1989 Optimization model for water distribution system design Journal of Hydraulic Engineering, ASCE 115 10 1401 1418

14. C. Xu, I. C. Goulter, 1998 Probabilistic model for water distribution reliability Journal of Water Resources Planning and Management, ASCE 124 4 218 228

15. Z. Kapelan, A. V. Babayan, D. Savic, G. A. Walters, S. T. Khu, 2004 Two new approaches for the stochastic least cost design of water distribution systems 4th World Water Congress: Innovation in Drinking Water Treatmen London

16. A. Babayan, D. Savic, G. Walters, 2005 Multiobjective optimization of water distribution system design under uncertain demand and pipe roughness School of Engineering, Computer Science and Mathematics, University of Exeter: Exeter

17. Z. Kapelan, D. Savic, G. Walters, 2005 An Efficient Sampling-based Approach for the Robust Rehabilitation of Water Distribution Systems under Correlated Nodal Demands School of Engineering, Computer Science and Mathematics, University of Exeter: Exeter

18. Y. Filion, B. Adams, A. Karney, 2007 Cross Correlation of Demands in Water Distribution Network Design Water Resources Planning and Management, ASCE 137 144

19. K. Pasha, 2005 Analysis of uncertainty on water distribution hydraulics and water quality Impacts of Global Climate Change Anchorage, Alaska

20. H. Tessendorff, 1972 Problems of peak demands and remedial measures Proc., 9th Cong. Int. Water Supply Assoc. int. standing committee on Distribution Problems: subject n.2

21. I. Rodriguez-Iturbe, V. K. Gupta, E. Waymire, 1984 Scale considerations in the modeling of temporal rainfall Water Resources Research 20 11 1611 1619

22. S. G. Buchberger, G. Nadimpalli, 2004 Leak estimation in water distribution systems by statistical analysis of flow reading Water Resources Planning and Management, ASCE 130 4 321 329

23. C. J. Hutton, L. S. Vamvakeridou-Lyrouda, Z. Kapelan, D. Savic, 2011 Uncertainty Quantification and Reduction in Urban Water Systems (UWS) Modelling: Evaluation report

24. O. Giustolisi, E. Todini, 2009 Pipe hydraulic resistance correction in WDN analysis Urban Water Journal 6 1 39 52

25. R. Magini, I. Pallavicini, R. Guercio, 2008 Spatial and temporal scaling properties of water demand Journal of Water Resources Planning and Management, ASCE 134 276 284

26. I. Rodriguez-Iturbe, M. Marani, P. D'Odorico, A. Rinaldo, 1998 On space-time scaling of cumulated rainfall fields Water Resources Research 34 12 3461 3469

27. E. Van Marcke, 1983 Random Fields Analysis and Synthesis (Revised and Expanded New Edition) World Scientific Publishing Co. Pte. Ltd

28. M. B. Fiering, 1964 Multivariate Technique for Synthetic Hydrology Journal of Hydraulics Division - Proceedings of the American Society of Civil Engineers 43 60

29. R. Guercio, R. Magini, I. Pallavicini, 2003 Temporal and spatial aggregation in modeling residential water demand Water Resources Management II UK: WIT Press

30. I. Pallavicini, R. Magini, 2007 Experimental analysis of residential water demand data: probabilistic estimation of peak coefficients at small time scales Proc. CCWI2007 & SUWM2007 Conf. Water Management Challenges in Global Change UK

31. O. Fujiwara, D. B. Khang, 1990 A two phase decomposition method for optimal design of looped water distribution networks Water Resources Research 26 4 539 549

32. L. A. Rossman, 2000 Epanet2 Users Manual Washington, USA: US EPA

Chapter 11

Optimal Upgradation and Expansion of Existing Water Distribution Networks Using Differential Evolution Algorithm

C.R. Suribabu and T.R. Neelakantan

Center for Advanced Research in Enviroment, School of Civil Engineering, SASTRA University, Thanjavur, 613 401, India

ABSTRACT

Most of water supply infrastructures were put in place during India's independence period and some pipelines are over 70 years old. Ageing and uncertainty on its performance under increased and varying demand pattern puts tremendous pressure on water authorities to rehabilitate the existing system for complete revamping the system. Complete revamping does not only mount the cost to the water authority but also makes potential inconvenience to the consumer in terms of getting new connection with new line. Hence, it is essential to rehabilitate the pipes which are more sensitive in serving the locality. Rehabilitating the existing system is generally done by either replacing the existing pipes with same or higher diameter or laying parallel pipe with existing. The optimal selection of such a combination is a furious task to water authorities. In the present study, a combined simulation-optimization modeling approach is developed using EPANET simulation engine. Differential evolution algorithm is adopted to arrive at the optimal solution. The demonstration proves that if the required data are available, the rehabilitation and expansion can be performed in an optimal manner using the model presented.

KEYWORDS

Water supply, rehabilition, simulation, optimization, differential evolution

1. INTRODUCTION

Generally, problems relating to water distribution networks fall under three broad categories; designing a new network, modifying or expanding an existing network and operating an existing network. On account of the unprecedented growth and development or due to the unexpected deterioration of the hydraulic capacity of pipes, a network may not perform at the desired level of service even within the service life and hence may have to be rehabilitated in order to improve its performance. Rehabilitation of an existing network could involve addition of new nodes and pipes, replacement of existing pipes with new ones, cleaning and lining of existing pipes, duplicating the existing pipes and addition of new pumps and tanks. Beyond the life time, repairing a pipe line is a short-term solution and generally carried out under strict budget constraints. The useful life of the network can be enhanced by replacing certain pipes and therein the structural strength of the pipes can be restored but at optimal cost. The optimal selection of such a combination is a furious task to water authorities.

Researchers have been investigating the cost effective water distribution network design optimization with various approaches such as linear, nonlinear, dynamic and mixed integer programming since 1970's. Goulter (1992) provided an extensive review of the optimisation models applied to water distribution design. Recently many evolutionary and heuristic optimization models (simulated annealing, genetic algorithm, ant colony algorithm, shuffled frog leap algorithm, particle swarm optimization and differential evolution) have been applied to water distribution network optimization problem (Simpson et al., 1994; Savic and Walters, 1997; Cunha and Sousa, 1999; Maier et al., 2003; Eusuff and Lansey, 2003; Suribabu and Neelakantan, 2006; Suribabu, 2010). For upgrading and enhancing existing water distribution networks, Boulos and Wood (1990) proposed a method for determining different parameters explicitly.

Rehabilitation and expansion of water distribution networks are generally treated as an optimization problem since it involves the sizing of pipe for parallel lines and also for new area as expansion to the existing area. Both traditional and non-traditional optimization approaches are documented by a number of researchers in the literature. The problem is generally formulated to minimize the cost of providing parallel lines and also to minimize the cost of pipes for expansion area where no account is made for planning time horizon. There are works which address the decision support framework for optimal time for replacement and rehabilitation that describes the general guidelines for deciding whether a given pipe should be replaced/rehabilitated over a given planning time horizon. Few works utilizes the life-cycle cost and performance measures for optimal design and rehabilitation of water distribution networks. Further, there are some works which considers the long-term strategic, operation-related replacement and rehabilitations.

For use in upgrading existing networks in order to sustain any single component failure, Ormsbee and Kessler (1990) suggested a linear programming-based method and Lingireddy et al. (2000) presented a genetic algorithm method. Lai and Schaake (1969) presented a methodology based on linear and dynamic programming for optimal expansion of New York City water supply problem; subsequently many other also applied different optimization techniques for optimal expansion of the same network. Stage-wise planning was demonstrated byMartin (1990) through a dynamic programming model to minimize the cost of initial construction and subsequent capacity expansion. Dandy and Engelhardt (2001) used a genetic algorithm to find an optimal schedule for the replacement of water supply pipes. Luong and Nagarur (2005)proposed a mathematical model to find an optimal fund allocation for better maintenance of distribution network and decision on repair or replacing the components in case of failure state. Dandy and Engelhardt (2006) developed trade off curves between cost and reliability for pipe replacement decisions for water distribution network of Northern suburbs of Adelaide, Australia. Jayaram and Srinivasan (2008) proposed a multi-objective formulation with “minimization of life-cycle cost” and “maximization of performance” as objectives for the optimal design and rehabilitation of a water distribution network. The life-cycle cost includes the initial cost of pipes, the cost of replacing old pipes with new ones, the cost of cleaning and lining existing pipes, the expected repair cost for pipe breaks and the salvage value of the pipes that are replaced. Neelakantan et al. (2008) proposed a methodology for the design of water distribution networks incorporating the pipe-break and the replacement cost economics.

Kettler and Goulter (1983), Goulter and Coals (1986), Su et al. (1987), Bao and Mays (1990), Fujiwara and Tung (1991), Cullinane et al. (1992), Prasad and Park (2004) and Suribabu and Neelakantan (2008) have developed reliability-based optimisation models. Despite the number of options for rehabilitation namely replacement, duplication, relining, cleaning and lining available, providing parallel pipes, replacing existing with new pipe and adding new overhead tanks are very common in developing countries. It is common that to add overhead tanks and providing parallel pipes to the existing water distribution network whenever demand increases and pressure decreases at consumer point.

Water distribution networks are generally designed for projected demand for design period but in reality, the demand at the middle or end of design period will not be matching with the projected value due to unexpected growth of urbanization in some part of distribution area. Hence, network needs improvement in terms of supplying capability and pressure management. This is generally accomplished by providing parallel pipes or adding elevated tanks to the network. This study demonstrates the optimal upgradation and expansion of existing water distribution system using differential evolution algorithm. The two benchmark networks and a real network are used for illustration. This study focuses on an expansion and upgradation procedure through a bi-level simulation-optimization model using a relatively new and simple heuristic optimization based differential evolution algorithm.

2. OPTIMIZATION MODEL

The topology of an urban water distribution system is generally imposed by the structure of the urban context: Roads, industrial areas, hospitals, etc. and a layout is generally available. The design demands in terms of quantity of water to be delivered are preliminarily established on the basis of per capita consumption and population growth, industrial development and fire hydrants, etc. Therefore, the optimal design of the water distribution to be solved assumes a predefined topology and known demands in order to find the combination of commercial pipe diameters that gives the least cost network by satisfying the constraints on pressures. The water distribution network design is formulated as a least-cost optimization problem while decision variables being duplicating the existing pipes or to replace the existing pipes with commercially available new pipe or to introduce a new parallel pipe.

Minimization of the total cost of the distribution system may be expressed mathematically as:

$$\text{Minimize cost} = C_{new} + C_{reh} \tag{1}$$

$$C_{new} = \sum_{i=1}^{N} Cp(D_i)L_i$$

$$C_{reh} = \sum_{r=1}^{R} C_{RH}(D_r)L_r$$

Where:

CP (DI) = Cost of per unit length of new pipe i with diameter DI

Ln = Length of new pipe i

CRH(Dr) = Cost of per unit length of rehabilitated pipe r with diameter Dr

Lr = Length of rehabilitated pipe r

This objective function needs to be minimized subject to a set of constraints as follows.

The first set of constraints depicts the law of conservation of mass at each node. The nodal conservation of mass requires that the sum of flows into or out of any junction node minus any external demand must be equal to zero. Application of nodal conservation of mass results in a set of linear equality constraints in terms of pipe flow rates as follows:

$$\sum_{i \in NP_{in,n}} Q_i - \sum_{i \in NP_{out,n}} Q_j = Q_{ext,n} \quad \forall n \in NN \tag{2}$$

Where:

Q = Pipe flow

Qext,n = Demand at node n

NPin,n = Set of pipes entering node n

NPout,n= Set of pipes leaving node n

NN = Node set

The second set of constraints depicts the energy conservation law in each loop or between every two nodes of known head. The head loss around the closed path (loop) should be equal to zero or the head loss along a path between the two fixed head nodes should be equal to the difference in elevation as follows:

$$\sum_{k \in loop_j} K_{li} Q_i^p = \Delta E_j \qquad \forall\ j \in NL \tag{3}$$

$$K_{li} = \frac{\alpha L_i}{C_{HW}^p D_i^{4.87}} \tag{3a}$$

Where:

Kli = Hydraulic resistance (function of diameter, material and roughness of pipes)

α = Co-efficient whose value depends on the units used

Li = Length of the link I

CHW = Hazen-Williams co-efficient

Di = Diameter of the Link I

NL = Loop set

ΔEj = The sum of energy losses around j loop that will be equal to 0 for closed loops and equal to the head difference between the ends of open loops p = exponent

The third set of constraints expresses the minimum pressure requirements at each node as follows:

$$H_n \geq H_{min} \tag{4}$$

The fourth set of constraints belongs to minimum and commercial set of diameters as follows:

$$D_i \geq D_{min} \quad \forall I \in NP \tag{5}$$

$$D_i \in [D] \quad \forall i \in NP \tag{6}$$

where, [D] = set of commercial diameters.

3. DIFFERENTIAL EVOLUTION ALGORITHM

Differential Evolution (DE) is a stochastic and population based search algorithm. DE utilizes the same operators of genetic algorithm and the main difference in constructing better solutions rely on mutation rather than cross-over operation. The evolution process starts with initial population containing individuals according to size of the population and gene values of each individual are generated randomly. The fitness value of each individual are calculated according to the fitness function. Three individuals are selected randomly from population and differences of genes of two individuals are calculated and it is multiplied by weighing factor called mutation (within the range of 0-1). The resulting weighted genes are added with third selected individual and a new individual is created by performing uniform crossover between the newly calculated gene values and the individual originally selected. The new individual's fitness is evaluated and compared to that of the original individual and fittest of the two individuals will be taken to the next generation. The above procedure is repeated for all individuals of a population and a new population is obtained. The entire procedure is repeated for a number of generations. The gene values of the best individual of the last generation are taken as the solution to the problem.

DE operates on a population P(G) of generation G that contains npop candidate solutions (individuals). The position of the population of generation G can be represented in the matrix format as below:

$$P^{(G)} = X_i^G \begin{bmatrix} x_{1,1}^G, x_{2,1}^G, x_{3,1}^G, \ldots\ldots, x_{npop,1}^G \\ x_{1,2}^G, x_{2,2}^G, x_{3,2}^G, \ldots\ldots, x_{npop,2}^G \\ x_{1,3}^G, x_{2,3}^G, x_{3,3}^G, \ldots\ldots, x_{npop,3}^G \\ . \\ . \\ . \\ x_{1,nvar}^G, x_{2,nvar}^G, x_{3,nvar}^G, \ldots\ldots, x_{npop,nvar}^G \end{bmatrix} \quad (7)$$

where, Xi(G) is the ith candidate solution or individual in the Gth generation, xi,j(G) is the jth component (diameter of the jth link) of the ith candidate solution, nvar denotes the number of variables (in this case, the number of diameters to be selected) and Gmax is the user-defined maximum number of generations.

The initial population is randomly generated within lower and upper bound of decision variables using following expression:

$$P_{(0)} = x_{i,j}^{(0)} = x_j^{(L)} + rand_{i,j}(x_j^{(U)} - x_j^{(L)}) \qquad i = 1 \text{ to } n_{pop},\; j = 1 \text{ to } n_{var} \quad (8)$$

where, 'randi, j' denotes a uniformly distributed random value within the range [0.0 to 1.0]. xj(U) and xj(L) are upper and lower limits of variable j.

From the first generation onwards, the population (new vectors) of the subsequent generation P(G+1) is generated by the combination of vectors randomly chosen from the current population by mutation. The noisy vector is then mixed with the predetermined target vector.

The population of 'trial' vectors P(G+1) is generated as follows (mutation and recombination):

$$x_{i,j}^{G+1} = \begin{cases} x_{C,j}^{(G)} + F \times (x_{A,j}^{(G)} - x_{B,j}^{(G)}) & \text{if } rand_{i,j} \leq C_r; \quad \forall j = D_i \\ x_{i,j}^{(G)} & \text{otherwise} \end{cases} \quad (9)$$

Where:

$D = 1, \ldots\ldots, n_{param}$

$A \in [1, \ldots\ldots, n_{pop}],\quad B \in [1, \ldots\ldots, n_{pop}],\quad C \in [1, \ldots\ldots, n_{pop}],\ A \neq B \neq C \neq i$

$C_r \in [0-1],\ F \in [0-1],\ rand \in [0-1]$

The weighting factor F is a user-defined mutation constant within the range [0-1]. Cr is a user-defined crossover constant which assists in the differential perturbation to select the parameters either from the noisy vector or target vector to get the trial vector. Finally, the trial vector is carried to the next generation only if it yields a reduction in the value of the objective function in the case of the minimization problem. Otherwise the target vector will be selected for the next generation.

The population of the next generation P(G+1) is selected as follows:

$$x_{i,j}^{G+1} = \begin{cases} x_{i,j}^{(G+1)} & \text{if } f(x_i^{(G+1)}) \leq f(x_i^{(G)}) \\ x_{i,j}^{(G)} & \text{otherwise} \end{cases} \quad (10)$$

where, f(xi(G)) is the cost of the ith individual in the Gth generation.

A detailed description of DE is available in Storn and Price (1995) and application to the water distribution network design is available inSuribabu (2010) and Vasan and Simonovic (2010).

4. APPLICATION

In the present study, a combined simulation-optimisation model is developed and used. The optimisation model is an outer driven whereas the simulation is an inner driven one. The computer programming code was written for DE using Visual Basic and EPANET (Rossman, 2000) is linked via., the EPANET Tool kit. The complete program performs a hydraulic network analysis at each function evaluation to determine the pressure head at the nodes. A penalty value will be added to the solution vector that violates pressure at the node. The penalty value is calculated if all the components forming with its maximum size. In the DE process, discrete sizes will be converted to continuous diameter when it undergoes the mutation process. From the initial population, two solution vectors are randomly selected and the difference between each of the parameters is determined. The weighted vector is obtained by multiplying the mutation constant and it is added to the third randomly selected vector from the initial population to get a noisy vector. Furthermore, a new solution vector is obtained by performing crossover which basically selects the pipe diameter either from the noisy vector or target vector according to the selection probability. The selection probability is randomly generated in order to compare with the crossover constant. If it is less than or equal to the crossover constant, the pipe diameter is selected from the noisy vector, otherwise from the target vector. The overall cost of the new solution (called a trial vector) is calculated after converting the pipe diameters selected from the noisy vector to the nearest commercial size. Since, conversion of continuous diameter to discrete is carried out before the selection of the vector for the next generation, the goal of optimisation will not be affected in the design process. All the decisions for selection of either parallel pipes or leaving as it is implemented using discrete numbering system in the optimization process.

Example network 1: The network shown in Fig. 1 consists of a reservoir, 17 nodes and 26 pipes. The elevation of each node is presented with the layout. Table 1 provides the length, diameter and roughness value for each pipe at the time of installation and also after 15 years (assumed).

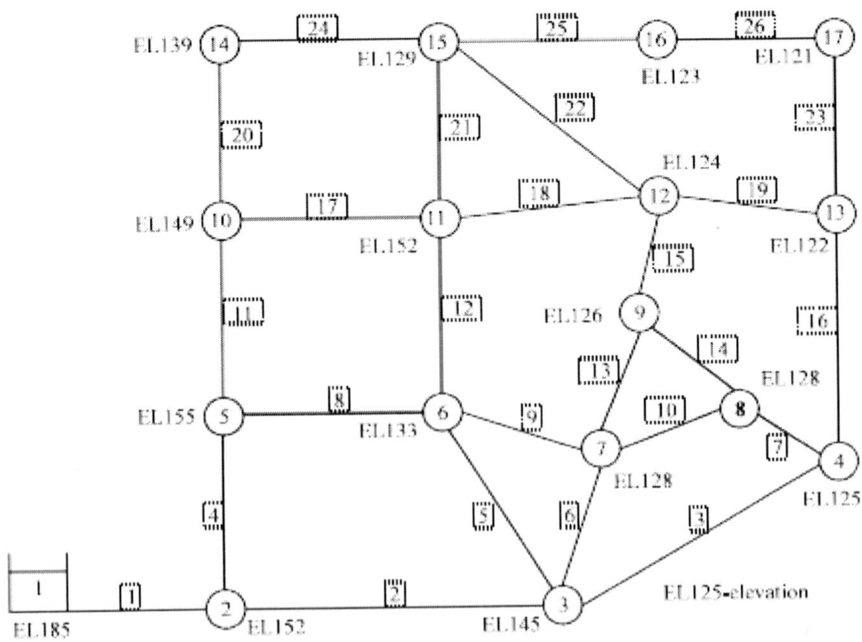

Figure 1: Layout of example network 1

Table 1: Network data for example network 1

Pipe No./node No.	Length (m)	Diameter (mm)	C_{HW} at the start of design period	C_{HW} after 15 years	Designed demand ($m^3 h^{-1}$)	Demand after 15 years ($m^3 h^{-1}$)
1	1030	406.4	130	100	Reservoir	Reservoir
2	890	254.0	130	98	50	70
3	1015	152.4	130	98	65	80
4	243	406.4	130	100	45	90
5	570	101.6	130	90	50	50
6	350	203.2	130	90	100	130
7	422	101.6	130	90	60	65
8	450	355.6	130	90	90	70
9	320	101.6	130	90	20	40
10	580	152.4	130	90	55	70
11	750	254.0	130	95	80	90
12	730	254.0	130	95	85	90
13	500	101.6	130	90	50	60
14	378	101.6	130	90	100	120
15	570	101.6	130	90	100	130
16	560	152.4	130	95	60	50
17	700	101.6	130	98	70	80
18	610	101.6	130	98		
19	631	101.6	130	95		
20	875	254	130	95		
21	890	101.6	130	95		
22	808	101.6	130	90		
23	826	152.5	130	90		
24	810	203.2	130	95		
25	585	152.4	130	90		
26	631	101.6	130	95		
					1080	1285

The network is actually designed for designed demand value shown in the Table 1. At the end of 15th year, the actual demand is estimated according to the urban development and future requirement and the projected demand is presented in the Table 2. Now, the network requires rehabilitation in order to supply increased demand resulted due to urbanization in the distribution area. While analysis of the network for new demand and revised roughness value (assumed), the pressure head at several nodes falls well below the minimum pressure of 15 m and some nodes experiences negative pressure. This shows that the existing pipe sizes could not supply the increased demand and change in roughness co-efficient resulted due to age of the pipes. Hence, the existing network needs rehabilitation in terms of replacing some of existing pipes with new pipes or to provide the parallel pipes. Three possible decisions are allotted for each pipe. The first option is to replace the existing pipe with new one, the second option is to provide parallel pipe along with existing pipe and last option being to the leave pipe as it is. The proposed simulation-optimization model is applied to evaluate decisions on rehabilitation. In this study, no energy cost is taken into account as the system taken for analysis is a gravity fed. Minimizing the capital cost of laying the new pipes is taken as the objective. Table 2shows the commercially available pipes and their cost per metre length. Thirty trial runs are performed with different initial random seeds, for each set of selected operator constants by setting a population size as 20. The mutation constant (weighting factor) is varied from 0.6-0.9 in 0.1 increments and similarly the crossover constant is varied from 0.3-0.5 in increments of 0.1. Ten different combinations of constant are considered from the above range. The termination criterion for the optimization is arbitrarily set to 5000 generations. As the population size is set to twenty, each generation consists of 20 function evaluations. Table 3 provides the least cost of the network obtained from 10 trial runs of each combination of constants.

The different random seed number for each trial run is maintained. It can be seen from the Table 3 that the least cost solution is obtained when weighting and crossover is 0.6-0.7 and 0.4, respectively. Further, using DE, the optimal cost for network in case of complete replacing with new pipes (no parallel pipes) is found out. To identify the least cost, 10 ten trial runs are performed for different combination DE operators. The least cost solution obtained for each combination of DE operators is presented in Table 3. The least cost solution for rehabilitation and complete replacement with new pipes were found out as Rs. 151, 38, 560, 312, 26 and 640, respectively. Table 4 shows optimal pipe sizes for both the cases. It is clear from the result that the optimal rehabilitation plan saves more than 50% of cost than optimal replacement of all the pipes by new pipe if it is adopted. It can be seen from the results that for rehabilitation plan, only one pipe is replaced by new one, eight parallel pipes are added and remaining pipes are left as it is. Optimization selects more pipes to be left as it is and also minimize the number of parallel pipes to be laid.

Table 2. Available pipe sizes and cost details

Sr. No.	Pipe diameter (mm)	Cost (Rs.)
1	101.6	765
2	152.4	1150
3	203.2	1665
4	254.0	2250
5	304.8	2910
6	355.6	3640
7	406.4	4460
8	457.2	5430
9	508.0	6875
10	558.8	7980
11	609.6	9100

Table 3: Results of trial runs for example network 1

Trial run No.	Weightage factor	Crossover probability	Cost (Rs. 1000 units) for rehabilitation	Cost (Rs. 1000 units) for completely new pipes
1	0.8	0.5	16133.12	31279.75
2	0.8	0.4	15806.82	31279.75
3	0.7	0.5	15763.99	31279.75
4	0.6	0.5	15138.56	31414.95
5	0.6	0.4	15597.63	31226.64
6	0.7	0.4	16257.96	31226.64
7	0.8	0.3	15806.82	31279.75
8	0.7	0.3	16257.96	31279.75
9	0.9	0.6	15942.97	31432.60
10	0.9	0.4	16064.32	31373.44

Table 4: Optimal solution for example network 1

Pipe ID	1	2	3	4	5	6	7	8	9
Rehab*	|| 355.6	Leave	|| 254	|| 304.8	Leave	Leave	Leave	Leave	|| 152.4
New*	508	203.2	152.4	457.2	101.6	101.6	101.6	457.2	203.2
Pipe ID	10	11	12	13	14	15	16	17	18
Rehab*	Leave	|| 203.2	Leave	Leave	Leave	Leave	Leave	Leave	Leave
New*	152.4	254	406.4	101.6	101.6	101.6	101.6	101.6	152.4
Pipe ID	19	20	21	22	23	24	25	26	
Rehab*	Replace with 304.8	||304.8	Leave	Leave	Leave	|| 203.2	Leave	|| 203.2	
New*	152.4	203.2	254	101.6	101.6	101.6	203.2	152.4	

Rehab*: Optimal rehabilitation plan, New*: Optimal replacement of all the pipes by new pipes, ||: Parallel pipe along with existing pipe (diameter in mm)

Example network 2: The second network shown in Fig. 2 is required to provide a pipeline for newly developed area in addition to satisfy the minimum pressure at the existing nodal points. In view of increased nodal demands, it is required to add new overhead tank either at any of the existing nodal point or new nodal point. The source for additional supply is available 500 m away from one of the node of existing system.

Figure 2 shows the layout of existing system with the layout of expansion area. Table 5 provides the nodal elevation, old demand and revised demand for meeting future needs. Table 6 gives the pipe length, diameter, roughness (assumed) of existing network. The cost of pumping main from source to various possible location of overhead tank is presented in Table 7. The bottom level of an overhead tank is fixed as 24 m above the respective nodal elevation.

The cost of overhead tank is not included in the optimization model since its cost is same whatsever the nodal location. The old and new hourly total peak demands are 1080 and 1675 m3, respectively. The additional demand of 595 m3 h-1 is to be drawn from new source. The expansion network is optimized without treating the location of new tank as an additional decision variable. Instead the expansion network is independently optimized for each possible location for tank. In order to keep the search space size minimum, the location of tank is dealt independently. In order to meet the revised demand and nodal pressure in the existing system, the option for providing parallel pipes to all existing pipelines are considered. Since, the network is larger than the previous example, the maximum number of generation is considered as 50,000. Several trial runs are attempted with different initial random seeds, for each set of selected operator constants by setting a population size as 20. The mutation constant (weighting factor) is varied from 0.6-0.9 in 0.1 increments and similarly the crossover constant is varied from 0.3-0.5 in increments of 0.1. Table 8 presents the results of the optimization. It can be seen from the result that adding overhead tank at node 13 results least cost. This node is nearest to the new source. It is obvious that tank nearest to source cost less as the length of pumping main is minimal. However, the selection of location for overhead tank also depends on the availability of open land for construction. Hence, the decision of tank location may be taken in consultation with field engineers. If the number of locations available to construct for new overhead tank is limited among the hydraulically feasible location for given tank height, then it is required to select from other alternatives. If the cost of pumping main is excluded in the optimization, then the addition of tank at node 18 results least cost network. Further, it is to be noted that number of parallel pipes required is less if the new tank is added in the expansion area.

Table 5: Nodal elevation, old and new demand particulars for example network 2

Node ID	Elevation	Old demand ($m^3 h^{-1}$)	New demand ($m^3 h^{-1}$)	Node ID	Elevation	Old demand ($m^3 h^{-1}$)	New demand ($m^3 h^{-1}$)
1	138	50	70	13	130	100	120
2	135	65	80	14	129	100	130
3	125	45	90	15	123	60	50
4	129	50	50	16	121	70	80
5	133	100	130	17	127	-	50
6	128	60	40	18	126	-	50
7	127	90	70	19	126	-	60
8	126	20	40	20	127	-	60
9	130	55	70	21	127	-	40
10	130	80	90	22	127	-	45
11	124	85	90	23	127	-	50
12	122	50	60	24	127	-	60

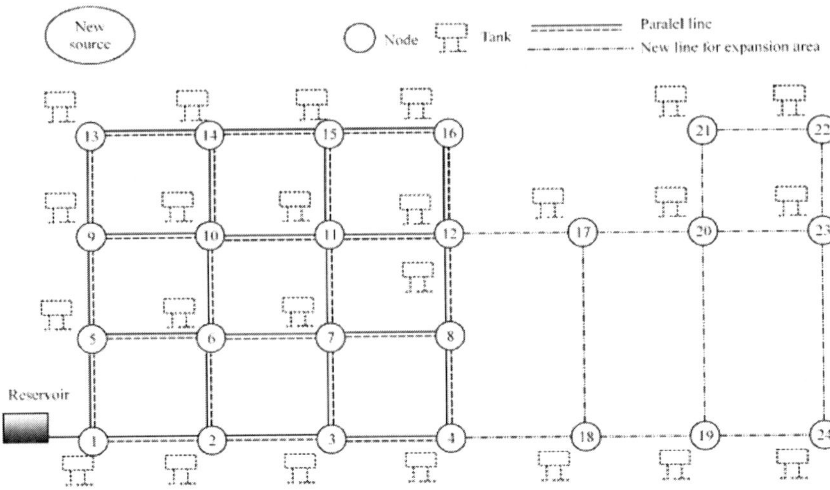

Figure2: Layout of example network 2

Example network 3 (real network): Tirumala (13°41′ N and 79°21′ E) in Andrapradesh State is one of the most popular pilgrimage centers in India. The temple town Tirumala is located on a hill at an altitude of 870 m above mean see level. Though the local population is very meager, the temple is visited by about 50000-500000 pilgrims daily. The maximum daily demand at Tirumala is about 17500 m3. The demand is mainly from guest houses, canteens, other public utilities and building constructions. Water is supplied to Tirumala by gravity from three reservoirs namely, Gogarbham, Akasaganga and Papavinasanam, located at higher altitudes in hills. Since, these sources are found insufficient during summer, water is being pumped from Kalyani reservoir (13°39′15″N and 79°16′20″) which is located downhill (about 700 m below Tirumala), at a rate of about 6100 m3 day-1. During summer days, water supply is rationed and about 2500 m3 day-1 of water is also fetched from the borewells.

Table 6: Length, diameter and roughness values for example network 2

Pipe ID (Node-node)	Length (m)	Diameter (mm)	C_{HW}	Pipe ID (Node-node)	Length (m)	Diameter (mm)	C_{HW}
1 (R-1)	1056	406	100	20 (10-14)	875	305	95
2 (1-2)	790	305	98	21 (11-15)	875	152	95
3 (2-3)	672	152	98	22 (12-16)	875	152	90
4 (3-4)	422	406	100	23 (13-14)	790	203	90
5 (1-5)	540	152	90	24 (14-15)	672	203	95
6 (2-6)	540	203	90	25 (15-16)	422	152	90
7 (3-7)	540	152	90	26 (12-17)	617	-	130
8 (4-8)	540	203	90	27 (4-18)	617	-	130
9 (5-6)	790	152	90	28 (17-18)	1255	-	130
10 (6-7)	672	152	90	29 (17-20)	560	-	130
11 (7-8)	422	406	95	30 (18-19)	560	-	130
12 (5-9)	750	203	95	31 (19-20)	1255	-	130
13 (6-10)	750	152	90	32 (20-21)	356	-	130
14 (7-11)	750	152	90	33 (21-22)	600	-	130
15 (8-12)	750	152	90	34 (22-23)	360	-	130
16 (9-10)	790	152	95	35 (20-23)	600	-	130
17 (10-11)	672	254	98	36 (23-24)	1255	-	130
18 (11-12)	422	152	98	37 (19-24)	600	-	130
19 (9-13)	875	152	95				

Table 7: Cost for providing pumping main from source to overhead tank

Node ID	Cost in (Rs. 10^3)	Node ID	Cost in (Rs. 10^3)	Node ID	Cost in (Rs. 10^3)
1	14500	9	7500	17	21100
2	18800	10	11800	18	28100
3	22500	11	15500	19	31000
4	25000	12	17750	20	24100
5	11600	13	2750	21	26100
6	15900	14	7000	22	29300
7	19500	15	10750	23	27500
8	21800	16	13000	24	34500

Table 8: Optimal solution for example network 2

Tank location	No. of parallel pipes required	Cost of new pipeline cost (in Rs. 10^3)	Optimal total cost (in Rs. 10^3)
1	9	28462.55	42962.55
2	5	25270.06	44070.06
3	7	28982.68	51482.68
4	3	20175.32	45175.32
5	10	33670.77	45270.77
6	6	25611.43	41511.43
7	6	22060.09	41560.09
8	4	20541.52	42841.52
9	4	18607.38	26107.38
10	2	13734.58	25534.58
11	4	18778.81	34278.81
12	5	22145.30	39895.30
13	6	22391.96	25141.96
14	3	20754.94	27754.94
15	4	22539.58	33289.58
16	5	28443.10	41443.10
17	3	15952.75	37052.75
18	3	11939.82	40039.82
19	3	13016.00	44016.00
20	3	24365.55	48465.55
21	4	19286.52	45386.52
22	3	27314.37	56614.37
23	3	22786.90	50286.90
24	3	14406.89	48906.89

Tirumala water distribution system is more than 50 years old. Since, the floating population is increasing every year, the water demand is also increasing. Many streets supported by single water main about 40 years ago now have 3 or 4 mains. Water from the above said sources is collected at a filter house in Tirumala and is supplied to the seven storage tanks. From these seven tanks, water is supplied through the water distribution networks. One of the seven water distribution networks, Gutta-II system is analyzed in this study. Gutta-II tank has a capacity of 788 m3 and is located at 941 m above mean sea level. Gutta-II water distribution network has been surveyed and details pertaining to the upgradation were acquired. Further actual demand was worked out based on the maximum possible population during peak festival season. The actual elevation of each node is found out using total station. Many parallel pipes of smaller size causes greater energy losses. Many of the parallel pipes are not functioning as they are heavily rusted and encrusted. In this situation, the water distribution network consisting of 96 existing links and 76 nodes, is used as a base network (Fig. 3). The optimization model is used for the upgradation of the network considering the options of parallel pipes, replacing the old links with new pipes and not changing the existing pipes. Minimizing the cost is the objective function while the minimum required pressures should be satisfied. The optimal result is presented inTable 9. It can seen from the Table 9 that optimization model selected 36 number of pipes is to be paralleled, 13 pipes have to be laid newly and 47 pipes have to be left as it is to meet out the designated demand with minimum pressure of 25 m at all the nodes.

5. CONCLUSION

Generally, water distribution system is the expensive part of the water supply system. The rehabilitation and expansion are very important for effective management of water distribution. However, the efforts on the long term planning of water distribution system including the rehabilitation and expansion seems very little as compared to the total budgets involved and to the potential savings thereof. Hence, this study is used to demonstrate the potential. A combined simulation-optimization modeling approach is used. For simulation, EPANET software is used while for optimization differential evolution, algorithm is used. The optimization algorithm uses the simulation iteratively number of times to arrive at the optimal solution. Since, EPANET has the facility to be called inside a computer programme where the parameters like diameter of a link can be changed, EPANET is used for simulation. Two hypothetical networks and a real network are used for demonstrating the model. The real network is taken from Tirumala Andrapradesh, India. The demonstration proves that if the required data are available, the rehabilitation and expansion can be performed in an optimal manner using the model presented.

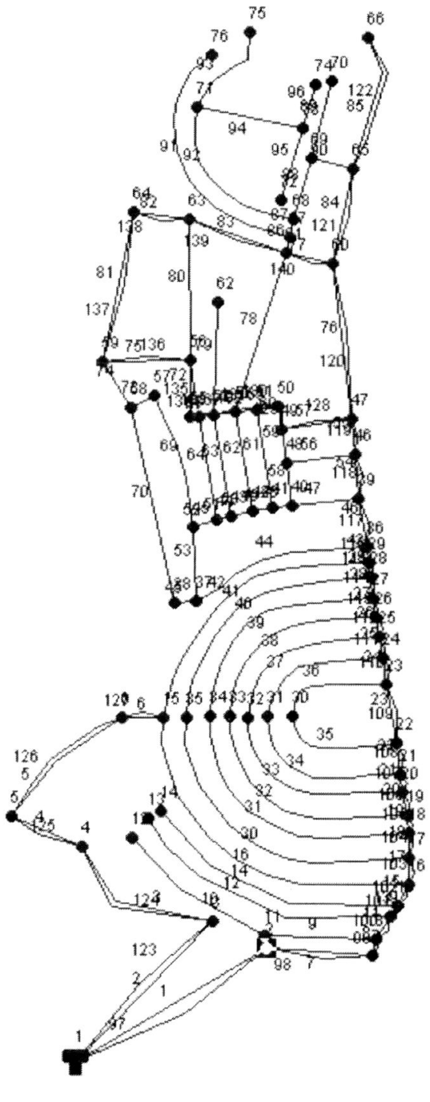

Figure 3: Layout of real network

Table 9: Results for real network

Link ID	Rehabilitation option	Diameter	Link ID	Rehabilitation option	Diameter
1	Parallel pipe	101.6	48	Leave	-
2	Parallel pipe	101.6	50	Leave	-
3	Parallel pipe	101.6	51	Leave	-
4	Parallel pipe	101.6	52	Leave	-
5	Parallel pipe	101.6	53	Leave	-
6	Parallel pipe	101.6	54	Parallel pipe	101.6
7	Parallel pipe	203.2	55	Parallel pipe	101.6
8	Parallel pipe	254.0	56	Leave	-
9	Leave	-	57	Leave	-
10	Leave	-	58	Leave	-
11	Parallel pipe	254.0	59	Leave	-
12	Leave	-	60	Leave	-
13	Parallel pipe	254.0	61	Leave	-
14	Leave	-	62	Leave	-
15	Parallel pipe	254.0	63	Leave	-
16	Leave	-	64	Leave	-
17	Parallel pipe	203.2	65	Leave	-
18	Parallel pipe	203.2	66	Leave	-
19	Parallel pipe	203.2	67	Leave	-
20	Parallel pipe	203.2	68	Leave	-
21	Parallel pipe	203.2	69	Leave	-
22	Parallel pipe	203.2	70	Parallel pipe	101.6
23	Parallel pipe	101.6	71	Leave	-
24	Parallel pipe	101.6	72	Leave	-
25	Parallel pipe	101.6	73	Leave	-
26	Parallel pipe	101.6	74	Parallel pipe	101.6
27	Parallel pipe	101.6	75	Leave	-
28	Parallel pipe	101.6	76	Leave	-
29	Parallel pipe	101.6	77	Leave	-
30	New	101.6	78	Leave	-
31	New	101.6	79	Leave	-
32	New	101.6	80	Leave	-
33	New	101.6	81	Parallel pipe	101.6
34	New	101.6	82	Parallel pipe	101.6
35	New	101.6	83	Parallel pipe	101.6
36	New	101.6	84	Parallel pipe	101.6
37	New	101.6	85	Parallel pipe	101.6
38	New	101.6	86	Leave	-
39	New	101.6	87	Leave	-
40	New	101.6	88	Leave	-
41	New	101.6	89	Leave	-
42	New	101.6	90	Leave	-
43	Parallel pipe	101.6	91	Leave	-
44	Parallel pipe	101.6	92	Leave	-
45	Leave	-	93	Leave	-
46	Parallel pipe	101.6	94	Leave	-
47	Leave	-	95	Leave	-
49	Leave	-	96	Leave	-

ACKNOWLEDGMENTS

The authors acknowledge the financial support of Department of Science and Technology, Government of India through their research grants

SR/S3/MCE/07/2002-SERC-Engg and DST/TSG/WP/2007/07. The authors thank SASTRA University for providing facilities for this study. The authors appreciate the cooperation extended by the Tirumala Tirupati Devestanam for giving permission to conduct the Total Station survey on the distribution system and providing other necessary information and data.

REFERENCES

1. Bao, T. and L.W. Mays, 1990. Model for water distribution system reliability. J. Hydraulic Eng., 116: 1119-1137.

2. Boulos, P.F. and D.J. Wood, 1990. Explicit calculation of pipe-network parameters. J. Hydraulic Eng., 116: 1329-1344.

3. Cullinane, M.J., K.E. Lansey and L.W. Mays, 1992. Optimization-availability-based design of water-distribution networks. J. Hydraulic Eng., 118: 420-441.

4. Cunha, M.C. and J. Sousa, 1999. Water distribution network design optimization: Simulated annealing approach. J. Water Resour. Plann. Mange., 125: 215-221.

5. Dandy, G.C. and M. Engelhardt, 2001. Optimal scheduling of water pipe replacement using genetic algorithms. J. Water Resour. Plann. Manage., 127: 214-223.

6. Dandy, G.C. and M.O. Engelhardt, 2006. Multi-objective trade-offs between cost and reliability in the replacement of water mains. J. Water Resour. Plann. Manage., 132: 79-88.

7. Eusuff, M.M. and K.E. Lansey, 2003. Optimization of water distribution network design using the shuffled frog leaping algorithm. J. Water Resour. Plann. Manage., 129: 210-225.

8. Fujiwara, O. and H.D. Tung, 1991. Reliability improvement for water distribution networks through increasing pipe size. Water Resour. Res., 27: 1395-1402.

9. Goulter, I.C. and A.V. Coals, 1986. Quantitative approaches to reliability assessment in pipe networks. J. Transport. Eng., 112: 287-301.

10. Goulter, I.C., 1992. Systems analysis in water-distribution network design: From theory to practice. J. Water Resour. Plann. Manage., 118: 238-248.

11. Jayaram, N. and K. Srinivasan, 2008. Performance-based optimal design and rehabilitation of water distribution networks using life cycle costing. Water Resour. Res., Vol. 44. 10.1029/2006WR005316

12. Kettler, A.J. and I.C. Goulter, 1983. Reliability consideration in the least cost design of looped water distribution systems. Proceedings of the 10th International Symposium on Urban Hydrology Hydraulics and Sediment Control, July 25-28, 1983, Lexington, Kentucky, pp: 305-312.

13. Lai, D. and J. Schaake, 1969. Linear programming and dynamic programming applications to water distribution network design. Report No. 116, Department of Civil Engineering, Massachusetts Institute of Technology, Cambridge, Massachusetts, USA.

14. Lingireddy, S., L.E. Ormsbee and T.R. Neelakantan, 2000. Application of genetic optimization in maximizing hydraulic network redundancy. Proceedings of the Joint Conference on Water Resources Engineering and Water Resources Planning and Management, July 30-August 2, 2000, Minneapolis, Minnesota, USA -.

15. Luong, H.T. and N.N. Nagarur, 2005. Optimal maintenance policy and fund allocation in water distribution networks. J. Water Resour. Plann. Manage., 131: 299-306.

16. Maier, H.R., A.R. Simpson, A.C. Zecchin, W.K. Foong, K.Y. Phang, H.Y. Seah and C.L. Tan, 2003. Ant colony optimization for design of water distribution systems. J. Water Resour. Plann. Manage., 129: 200-209.

17. Martin, Q.W., 1990. Linear water-supply pipeline capacity expansion model. J. Hydraulic Eng., 116: 675-690.

18. Neelakantan, T.R., C.R. Suribabu and S. Lingireddy, 2008. Optimisation procedure for pipe-sizing with break-repair and replacement economics. Water SA, 34: 217-224.

19. Ormsbee, L. and A. Kessler, 1990. Optimal upgrading of hydraulic-network reliability. J. Water Resour. Plan. Manage., 116: 784-802.

20. Prasad, T.D. and N. Park, 2004. Multiobjective genetic algorithms for design of water distribution networks. J. Water Resour. Plann. Manage., 130: 73-82.

21. Rossman, L.A., 2000. EPANET 2-user manual. National Risk Management Research Laboratory, Office of Research and Development, U.S. Environmental Protection Agency, Cincinnati, Ohio, USA.

22. Savic, D.A. and G.A. Walters, 1997. Genetic algorithms for least-cost design of water distribution networks. J. Water Resour. Plann. Manage., 123: 67-77.

23. Simpson, A.R., G.C. Dandy and L.J. Murphy, 1994. Genetic algorithms compared to other techniques for pipe optimization. J. Water Resour. Plann. Manage., 120: 423-443.

24. Storn, R. and K.V. Price, 1995. Differential evolution: A simple and efficient adaptive scheme for global optimization over continuous spaces. Technical Report, International Computer Science Institute, Berkley, CA., USA.

25. Su, Y.C., L.W. Mays, N. Duan and K.E. Lansey, 1987. Reliability-based optimization model for water distribution systems. J. Hydraulic Eng., 114: 1539-1556.

26. Suribabu, C.R. and T.R. Neelakantan, 2006. Design of water distribution networks using particle swarm optimization. Urban Water J., 3: 111-120.

27. Suribabu, C.R. and T.R. Neelakantan, 2008. Reliability based optimal design of water distribution networks by genetic algorithm. J. Intell. Syst., 17: 143-156.

28. Suribabu, C.R., 2010. Differential evolution algorithm for optimal design of water distribution networks. J. Hydroinformatics, 12: 66-82.

29. Vasan, A. and S.P. Simonovic, 2010. Optimization of water distribution network design using differential evolution. J. Water Resour. Plann. Manage., 136: 279-287.

Index